PROCEEDINGS OF THE INTERNATIONAL SYMPOSIUM ON
METALS AND THE ENVIRONMENT
Montréal, Québec, Canada, May 5 - 6, 1998

Metals and the Environment

Editor

J.M. Skeaff
Environment Laboratory, Mining and Mineral Sciences Laboratories
CANMET, Natural Resources Canada
Ottawa

Symposium organised and sponsored by The Metallurgical Society of CIM in celebration of the 100th Anniversary of The Canadian Institute of Mining, Metallurgy and Petroleum

CIM ANNUAL GENERAL MEETING MAY 4 - 8, 1998
Metals and the Environment

Metals and the Environment

3400 de Maisonneuve Blvd. W., Suite 1210
Montréal, Québec, Canada H3Z 3B8

Revised June 1998

ISBN: 0-919086-78-0
Printed in Canada

Foreword

The International Symposium on Metals and the Environment was organised and sponsored by The Metallurgical Society of CIM as MetSoc's contribution to the 100th anniversary celebrations of the Canadian Institute of Mining, Metallurgy and Petroleum held during the CIM Annual General Meeting in Montréal, May 4 - 8, 1998.

The purpose of the Symposium was to provide a forum for the review, discussion and informed debate of the environmental issues related to the metals industries, from both national and international perspectives. The Symposium was organised into four one-half day sessions:

- Session 1: National and International Environmental Protection Measures -- An Operational Framework for the Metal Producing Industries;
- Session 2: Dealing with Releases from Metal Industries;
- Session 3: Metal Commodities and Products; and
- Session 4: Resolving Metals Issues -- The Way Ahead

The Metallurgical Society considered that the Symposium would provide a timely opportunity to discuss and debate the very considerable challenges facing the metals industries as we head into the 21st Century. Therefore, the *ad hoc* Symposium organising committee invited panelists whom they believed would present to the CIM a wide spectrum of views of importance to the metals industry. Hence Session 1 was intended to ensure that the environmental protection framework in which the metals industries operates is well-understood. In Session 2, panelists endeavoured to place trace metal releases to the environment into the perspective of natural and anthropogenic activities and events. Session 3 panelists discussed the most arguably significant challenge facing the metals industries: continued open access to markets. A path forward for the metals industries, outlining important areas of future focus and effort, was the subject of Session 4.

The *ad hoc* Symposium committee would like to thank the panelists for preparing the papers for this Proceedings volume, and sincerely hope that the panelists' viewpoints will bring improved awareness of the key environmental issues for the metals industries.

I should like to express my thanks to the *ad hoc* Symposium committee for their very hard work in organising the technical program and the many other aspects of the Symposium. Special thanks are due to MetSoc, Environment Canada - Ontario Region, the International Council on Metals and the Environment, and Natural Resources Canada for supporting the Symposium.

Jim Skeaff

Montréal
May 1998

Editor's Biography

Jim Skeaff was born in Toronto and graduated in 1971 from the University of Toronto with a Ph.D. from the Department of Metallurgy and Materials Science. After investigating thermochemical equilibria in solid state metal-sulphur-oxygen systems at the University of Trondheim in Norway and corrosion products in the sodium-sulphur battery at Rensselaer Polytechnique Institute in Troy, N.Y., he joined the Mines Branch of the Department of Energy, Mines and Resources in Ottawa in 1974. While with the Department, he has investigated alternative methods for uranium extraction, conducted electric furnace smelting studies, determined the thermochemistry of arsenides and arsenates, and worked on the development of sensors for pyrometallurgical processes. More recently, he has become involved in national and international environmental issues related to the metal-producing industries, including the cooperative development of a laboratory protocol for the hazard identification of metals and inorganic metal compounds with the OECD.

Session Chairmen

National and International Environmental Protection Measures: An Operational Framework for the Metal Industries
Chair:
Gordon Peeling
Mining Association of Canada
Ottawa, Canada

Vice-Chair
Bill Blakeman
Environment Canada
Hull, Canada

Dealing with Releases from Metal Industries: Sources of Metals to the Environment Released During Production and/or Combustion
Chair:
Steve Hart
Environment Canada
Hull, Canada

Vice-Chairs:
Patrick Chevalier and Aleksander Ignatow
Natural Resources Canada
Ottawa, Canada

Metal Commodities and Products
Chair:
Gary Nash
International Council on Metals and the Environment
Ottawa, Canada

Vice-Chairs:
Len Surges and Hennie Veldhuizen
Noranda Mining & Exploration Inc.
Toronto, Canada

Resolving Metals Issues -- The Way Ahead
Chair:
Larry Seeley
Lakefield Research Limited
Lakefield, Ontario, Canada

Vice-Chair:
Mike Sudbury
Falconbridge Limited
Toronto, Canada

Contents

National and international environmental protection measures: An operational framework for the metal producing industries

The Minerals and Metals Policy of the Government of Canada: Partnerships for Sustainable Development

Linda J. Keen
Assistant Deputy Minister
Minerals and Metals Sector, Natural Resources Canada
580 Booth St., Ottawa Canada K1A 0E4

ABSTRACT

Canada is among the foremost producers and is the leading exporter of minerals and metals in the world. The industry plays an essential role in Canada's economy and provides jobs for more than 340,000 Canadians. It is the economic foundation for some 150 communities in Canada's rural and northern regions and plays an important role in the economic and social fabric of Canadian society. Canada's minerals and metals industry is facing a number of changes brought on by concerns about the state of our natural environment, increased competitive forces in the global economy, the need for a more efficient and effective federation, and the need to achieve sustainable development. The Minerals and Metals Policy addresses these challenges and responds to important government commitments.

The Policy sets parameters for federal decisions on minerals and metals, including a recognition of the need for sound science, the importance of global markets, and the benefits of both regulatory and non-regulatory approaches in pursuing environmental protection. It also commits the Government to continued efforts to secure international mineral investment for Canada and notes the importance of a fiscal and tax environment that recognizes the special challenges of the industry and its global context. The Policy also outlines the Government's support of the safe-use principle for minerals and metals, and engages the Government to advance this principle both domestically and internationally.

The Government is committed to a prosperous Canadian minerals and metals industry. The Minerals and Metals Policy provides the Government with the direction and decision-making tools that will help to ensure that the industry remains viable and responsible into the 21st century and beyond.

Proceedings of
Metals and the Environment: An International Symposium
Edited by Jim Skeaff
The Metallurgical Society of CIM
100th CIM Annual General Meeting, 1998

INTRODUCTION

The Minerals and Metals Policy of the Government of Canada: Partnerships for Sustainable Development describes, within areas of federal jurisdiction, the Government's role, objectives and strategies for the sustainable development of Canada's mineral and metal resources. The policy builds on important commitments and initiatives of the Government including Creating Opportunity, the Government's Mining Agenda, A Guide to Green Government, the Toxic Substances Management Policy, and the principles and goals of the Whitehorse Mining Initiative (WMI) Leadership Council Accord. It does this by contributing to three key elements of the Government's agenda: promoting economic growth and job creation, furthering an efficient and effective federation, and meeting the challenge of sustainable development.

Minerals and metals are of vital interest to Canada and are relevant to federal policies and programs because of their substantial contribution to Canada's social and economic well-being. Two important and interrelated developments have implications for Canada: the globalization of the industry, and the mounting need for governments around the world to collaborate in the development of solutions to environmental concerns and other challenges.

Provincial governments in Canada are responsible for mining within their respective jurisdictions. In this context, the federal government's role in minerals and metals has been more sharply focused on core federal responsibilities, including international trade and investment, science and technology, environmental protection, and Aboriginal affairs. The Government is committed to forging enhanced partnerships with others in exercising its responsibilities in minerals and metals.

The Government has adopted the Brundtland Commission's definition of sustainable development. The policy applies this definition by identifying the key elements of sustainable development in the context of minerals and metals.

In light of the foregoing, the policy has six major objectives:

- integrating the concept of sustainable development in federal decision-making affecting the minerals and metals industry;

- ensuring the international competitiveness of Canada's minerals and metals industry in the context of an open and liberal global trade and investment framework;

- advancing the concept of sustainable development of minerals and metals at the international level through partnerships with other countries, stakeholders, and multilateral institutions and organizations;

- establishing Canada as a global leader in promoting the safe use of minerals and metals, and their related products;

- promoting Aboriginal involvement in minerals- and metals-related activities; and

- providing a framework for the development and application of science and technology to enhance the industry's competitiveness and environmental stewardship.

FEDERAL DECISIONS IN MINERALS AND METALS: IMPLEMENTING A SUSTAINABLE DEVELOPMENT APPROACH

To achieve sustainable development, environmental, economic and social considerations must be taken into account as early as possible in the decision-making process. To help the Government meet this challenge in the area of minerals and metals, the policy enunciates a number of principles for sustainable development-based decision-making, including:

- a responsive public policy framework;
- the role of the market mechanism;
- the role of regulation;
- the role of non-regulatory approaches;
- the importance of science;
- endorsement of the concept of pollution prevention;
- affirmation of the precautionary principle; and
- recognition of the polluter pays principle.

THE BUSINESS CLIMATE: ENSURING THE COMPETITIVENESS OF CANADA'S MINERALS AND METALS INDUSTRY

Canada must compete as never before to attract investment capital to sustain its minerals and metals industry. In this environment, all governments must work together to ensure that a positive investment climate is maintained through i) policy, regulatory and fiscal reforms, and ii) the effective promotion of the improvements achieved through those reforms. Significant progress has been made by the federal government in both regards. Nevertheless, room for improvement remains. To address the need to inform investors, the Government will continue to promote Canada as a leading source of mineral and metal commodities, minerals- and metals-related semi-fabricated and fabricated products, and minerals- and metals-related technology and services, including technology and consulting services in such fields as mineral exploration, environmental protection and management. It will undertake these activities in close collaboration with the industry and provincial and territorial governments, utilizing the "Team Canada" approach.

Canada has become one of the world's principal venture capital markets for mineral exploration and development. Obstacles remain, however, to the ready access by firms to risk capital. Consequently, the Government supports, in partnership with interested provinces, the reduction of the multiple securities-related regulatory requirements for raising investment funds through the development of a Canadian Securities Commission and other appropriate mechanisms. Such action would help make securities regulations and capital markets across Canada more coordinated and efficient, thus facilitating the raising of risk capital, especially for the junior sector. The Government's fiscal regime is an important component in promoting investment, maintaining competitiveness and ensuring a positive investment climate. The Government has introduced many important fiscal improvements. These have been based on the principles of fairness, simplicity, economic growth, stability of revenues, and competitiveness. In developing new fiscal measures, and in recognition of the important role that the minerals and metals industry plays in the Canadian economy, the Government will:

- afford the minerals and metals industry fiscal treatment that recognizes exploration risk, ore reserve risk, and other specific risks unique to the sector;

- provide, to the extent possible, fiscal treatment to the Canadian industry that is competitive with the type of treatment afforded by other governments to mineral developers and producers operating in their jurisdiction;

- seek to ensure that government measures aimed at cost recovery reflect the cost of providing the goods or services received. Such measures should also be developed in consultation with industry and take into account competitiveness concerns; and

- within areas of federal jurisdiction, maximize the benefits for Canadians of their mineral resource endowment by ensuring that royalty rates and mining taxes are set at an equitable level, taking into account the industry's need to realize a return on its investment that is reflective of the risk taken and opportunity costs involved, and bearing in mind that Canada's mineral endowment is largely a publicly owned resource.

The Government also sets out a seven-item checklist for the development of any new federal regulatory processes that affect minerals and metals. As well, the policy states that the industry must continue to assume greater responsibility for environmental performance and for stewardship of minerals and metals throughout their life cycle.

MINERALS, METALS AND SOCIETY: PROMOTING PRODUCTS, MARKETS AND STEWARDSHIP

Minerals and metals are naturally-occurring substances that are indispensable materials for the development of human societies. While it is recognized that trace amounts of a number of minerals and metals are essential to all life forms, the extraction and processing, as well as the particular uses, of certain minerals and metals can and have resulted in adverse effects on human health and the environment. The federal government is committed to mitigating these effects through an improved scientific understanding of the role and behaviour of these substances, through the implementation of sound risk reduction and risk management strategies, and through the administration of its jurisdictional responsibilities concerning human health and the environment. Given Canada's role as a world leader in the production of minerals and metals, the management of issues related to health and the environment is a policy priority.

In managing minerals- and metals-related health and environmental issues, the principle of life-cycle management for both process and product life cycles plays an essential role. Process life-cycle management applies to specific operations and their associated risks in relation to the production of minerals and metals, such as exploration, extraction, processing, smelting and refining. It includes waste management, decommissioning and site rehabilitation. Product life-cycle management applies to specific elements, substances or products and their associated risks based on assessments of all stages in the cycle of manufacturing, use, re-use, recycling and disposal of that particular element, substance or product.

Life-cycle management of minerals and metals involves the application of risk-assessment and risk-management approaches: Risk assessment estimates the degree and likelihood of adverse effects resulting from exposure to a substance from a process or product. Risk management is the process of deciding what to do about an assessed risk, taking into account the results of the assessment as well as legal, economic and social factors.

The safe-use principle takes a life cycle-based approach to the use and management of minerals and metals, including the application of risk assessment and risk management strategies, in accordance with well-established stewardship practices. Borrowed from occupational health and safety terminology and focused on use, the safe-use principle integrates the notion of "risk" with "use." The principle builds on and complements the federal Toxic Substances Management Policy (TSMP). In doing so, it sends the message domestically and internationally that minerals and metals and their products can be used safely and responsibly.

In practical terms, the safe-use principle calls for an assessment of the risks at various stages throughout the life cycle associated with the uses of a mineral- or metal-containing product. The results of the assessment are then used to determine the most appropriate management approaches to address these risks. It is generally accepted that, in some cases, the risks associated with certain products or product uses cannot be properly controlled or managed. Consequently, where such a situation exists, the Government will either discontinue or prohibit the specific product or product use. In implementing the safe-use principle, and in the context of the TSMP, the Government will take the following approaches both domestically and internationally. The Government will:

- work with Canada's minerals and metals industry to develop and implement strategies that promote the safe-use principle. Such strategies would include, for example, the adoption of stewardship programs based on the life-cycle approach. The industry can and must play a leadership role in addressing health and environmental issues related to minerals and metals;

- ensure that Canada is a global leader in promoting the safe-use principle for environmental and health initiatives of relevance to minerals, metals and their products; and

- promote effective mechanisms to obtain and coordinate input from the provincial and territorial governments and others for the development of appropriate policy responses to scientific or technological advances in the safe management of minerals and metals and related products.

Recycled minerals and metals constitute an important source of secondary materials for industry and generate environmental benefits. As a consequence, the Government will work to: enhance the efficiency and effectiveness of regulations; promote a more efficient metals recycling industry in Canada; advance recycling as a feature of product design; and, at the international and domestic levels, promote common approaches to the definition of waste (including a distinction between metal-bearing recyclables destined for recovery and wastes destined for final disposal).

The federal government has a role to play in the reclamation of mine sites within its areas of responsibility, including establishing fiscal and regulatory conditions respecting reclamation for mine development on federal lands. The policy recognizes the need to clean up those abandoned and orphaned mine sites within federal jurisdiction that represent an unacceptable risk to the environment or human health and safety. It also acknowledges the need for site owners, where they can be identified, to pay for clean-up costs.

Land access for mineral exploration and development is necessary if the minerals and metals industry is to continue to contribute to Canada's economic and social well-being. In regard to Canada's ocean territory, that access will be determined through an integrated oceans governance strategy adopted by the Government. In addition to land access, governments must provide reasonable certainty to the industry that when it finds a mineral deposit, it can develop that deposit.

The Government affirms its commitments respecting the completion of the National Parks network and the establishment of National Marine Conservation Areas. It also remains committed to identifying and protecting terrestrial and marine critical wildlife habitat in Canada, and developing and implementing protected area strategies for federal lands and waters. In meeting these commitments, the Government will follow certain guidelines that recognize the important economic and social role played by the minerals and metals industry in Canada.

ABORIGINAL COMMUNITIES: PROMOTING INVOLVEMENT IN MINERALS AND METALS ACTIVITIES

Aboriginal concerns and interests in relation to mineral development are important factors in the Government's policy approach. The Government also respects existing provincial, territorial and municipal mechanisms for mineral development. Within matters of federal jurisdiction, it promotes cost-effective regimes for the sustainable development of minerals and metals on lands under claim, settlement areas, and Indian reserves.

The Government affirms its support for the timely resolution of land claims to remove uncertainty over the ownership and use of land and resources, as well as to encourage self-reliance by Aboriginal communities and promote their participation in economic opportunities.

The Government acknowledges Aboriginal concerns about the effect of mineral exploration and development on traditional lifestyles and the environment, as well as the desire of Aboriginal peoples to be involved in decision-making. Collaboration between the industry and Aboriginal communities is encouraged. The Government supports partnership approaches involving Aboriginal communities and the industry.

SCIENCE AND TECHNOLOGY: PROGRESS THROUGH INNOVATION

Science and technology (S&T) play a critical role in the health and well-being of Canadians, our ability to protect the environment, and our success in creating jobs and fostering economic growth. Federal S&T activities that are based on associated goals (i.e. job creation and economic growth, improved quality of life, and the advancement of knowledge) will continue to support sustainable development objectives. In addition to focusing on these goals in an integrated manner, the Government supports the development of stronger science-policy linkages.

In this light, the Government is committed to pursuing a number of goals related to Canada's S&T activities in relation to minerals and metals. It also affirms its commitment to promote partnerships among stakeholders in pursuing these goals.

The Government will pursue the following strategic, long-term directions in S&T related to minerals and metals:

- providing a comprehensive geoscience information infrastructure;

- supporting a sustainable minerals and metals industry (through the use of S&T to promote technological innovation both in mining operations and in the safe and efficient use of minerals and metals);
- enhancing the health and safety, and environment quality of Canadians;
- promoting the competitiveness of the Canadian industry; and
- developing value-added mineral and metal products.

MINERALS AND METALS AT THE INTERNATIONAL LEVEL: PROVIDING LEADERSHIP IN THE IMPLEMENTATION OF SUSTAINABLE DEVELOPMENT

Canada plays a leadership role at the international level, deriving from its position as the world's largest exporter of minerals and metals, and is a major player in the promotion of sustainable development, including the implementation of the Rio Summit's Agenda 21. The international nature of many of the pressures on the sector and the lessons learned to date, including the potential of initiatives aimed at environmental, health and social concerns to affect the competitiveness and acceptability of minerals and metals in the marketplace, requires an effective and flexible response by the Government.

The Government reaffirms its commitment to a liberalized, predictable, rules-based international trading and investment regime. It also sets out objectives and conditions for the promotion of the sound management of minerals and metals through a variety of instruments from legally-binding agreements to government-sponsored non-regulatory approaches to voluntary industry initiatives. Central to the Government's approach are the concepts of risk assessment, risk management, and the application of the safe-use principle. The Government reaffirms the value of bilateral and regional cooperation, and makes a commitment to continue to provide technical cooperation based on Canada's solid foundation of minerals- and metals-related knowledge, expertise and world-class technology.

MEASUREMENT AND FOLLOW-UP

The Government recognizes that criteria and indicators are essential because they help us to measure progress by industry toward the achievement of sustainable development. Accordingly, the Government views the development of sustainability criteria related to environmental, economic and social objectives and indicators for minerals and metals as a priority.

The policy recognizes the need for ongoing accountability for, and assessment of, results that flow from the policy. To this end, the Minister of Natural Resources, in cooperation with other federal departments and agencies, will issue periodic progress reports on its implementation. Public accountability for the implementation of the policy will also be ensured through the Commissioner of the Environment and Sustainable Development in the Office of the Auditor General.

In implementing this policy, the Government will take into account appropriate internationally established standards and Canada's international commitments to sustainable development and to free trade. By working bilaterally and multilaterally with other countries, and through international fora such as the United Nations and its agencies, the Orgnaisation for Economic Cooperation and Development (OECD), and the World Trade Organization (WTO), Canada will promote the policy's objectives and strategies, including the sustainable development of minerals and metals, both domestically and internationally.

Protecting the environment: An overview of Canada's policy, assessment and regulatory initiatives

Karen M. Lloyd
Commercial Chemicals Evaluation Branch, Environment Canada
351 boul. St. Joseph, Hull Canada

ABSTRACT

This presentation addresses several policy, assessment and regulatory initiatives in which Environment Canada is leading or involved, both nationally and internationally, dealing with minerals and metals. The framework for the risk management of substances in Canada is provided by the federal government's Toxic Substances Management Policy. Application of this policy to naturally occurring substances such as metals will be discussed. Environment Canada's role in international activities, such as those within the OECD, UN-ECE, and the NAFTA/CEC will also be described. Goals of some continental initiatives, such as those focused on the Great Lakes, will be briefly mentioned. Nationally, a description of the Priority Substances Assessment program of the Canadian Environmental Protection Act will be provided, with particular emphasis on the assessments of releases from primary and secondary copper and zinc smelters and refineries. Links of assessment and risk management activities to the research community will be identified, using the PSL program as an example.

Proceedings of
Metals and the Environment: An International Symposium
Edited by Jim Skeaff
The Metallurgical Society of CIM
100th CIM Annual General Meeting, 1998

INTRODUCTION

Environment Canada is a science-based department whose business is helping Canadians live and prosper in an environment that is properly protected and conserved. To address the risk posed to the environment or to human health in Canada from chemicals in the Canadian environment, Environment Canada is leading or is involved in a number of initiatives. These initiatives at the international, continental and national levels include policy, assessment and regulatory programs which have a strong link to the scientific community. The purpose of this presentation is to give a brief overview of some of the programs at the international, continental and national levels, which are of interest to the metals and minerals sectors. While this is not an exhaustive list of initiatives, the list should highlight our approach to protecting the Canadian environment.

FRAMEWORK

Toxic Substances Management Policy

The federal government's Toxic Substances Management Policy (TSMP), adopted in June 1995, provides the federal framework for making science-based decisions on the effective management of toxic substances. It guides federal actions domestically and serves as the centrepiece for the Canadian position in international negotiations on managing toxic substances. It applies to areas within federal jurisdiction.

The policy guides federal regulatory and non-regulatory programs by defining the ultimate management objective for a substance. The policy has two main management objectives:

1. virtual elimination from the environment of toxic substances that result primarily from human activity and that are persistent and bioaccumulative (Track 1 substances), and
2. management of other toxic substances and substances of concern throughout their life cycles to prevent or minimize their release into the environment (Track 2 substances).

The criteria for determining whether a substance is managed under Track 1 or Track 2 are presented in Table 1. A substance that meets all four criteria, in other words, that is persistent, bioaccumulative, toxic and primarily the result of human activity, will be targeted for virtual elimination from the environment (Track 1). This objective will be achieved by addressing sources of release to the environment or by removing or managing the substance if it is already in the environment.

TABLE 1. Criteria used to determine Track 1 Substances in the TSMP.

Persistence	Bioaccumulation	Toxicity	Predominantly Anthropogenic
Medium Half-life Air $\geq$ 2 days Water $\geq$ 182 days Sediment $\geq$ 365 days Soil $\geq$ 182 days	BAF $\geq$ 5,000 or BCF $\geq$ 5,000 or log $K_{ow} \geq 5.0$	CEPA-toxic or CEPA-toxic Equivalent	Concentration in environment largely resulting from human activity

Substances that do not satisfy all four criteria in Table 1 are managed under Track 2 and are candidates for full life-cycle management to prevent or minimize their release into the environment. A substance's source of release is a fundamental consideration in selecting risk management strategies. Some substances that are persistent, bioaccumulative and toxic occur naturally and can never be eliminated from the environment. Elements and naturally occurring inorganic compounds can fit into this category, and as such are not candidates for Track 1. Elements and naturally occurring substances that are used or released as a result of human activity may be targeted under Track 2 for reduction to naturally occurring levels, which would vary from region to region within Canada.

Since toxic substances or substances of concern can originate either within Canada or abroad, domestic actions have to be complemented by international measures to protect the Canadian environment. As Canada takes a leadership role in seeking international action, this policy serves as the centrepiece for the country's position on managing toxic substances in discussions and negotiations with the world community.

INTERNATIONAL INITIATIVES

Heavy Metals Protocol

Canada is a party to the 54 member United Nations Economic Council for Europe (UN ECE) and through this program we have signed onto the Convention on Long Range Transboundary Air Pollution (LRTAP). A recent initiative under this program is the development of a Protocol to control air emissions of heavy metals. Negotiation of this Heavy Metals Protocol began in March 1997 and is expected to be completed at the third negotiation session in January 1998 with signing anticipated in June 1998. The Protocol will then be ratified following confirmation of domestic harmonization through the Canada Wide Standards program of the Canadian Council of Ministers of the Environment (CCME).

The Heavy Metals Protocol includes cadmium, lead and mercury. While there is some provision under the protocol for products, the only mandatory product control will be that pertaining to the use of lead in gasoline, which is still in use in some of the UN ECE nations. Any other product inclusions will be recommendatory only, unless they are shown to have direct and significant impacts through long range air transport. Control options for major stationary sources will include percent caps or reductions by a target year for existing facilities and emission limit values and best available techniques for new facilities. These options are complimentary to currently existing Canadian regulations. Other aspects of the protocol include: a focus on the capture of particulates as the major control strategy; consideration for the use of non-regulatory mechanisms to achieve equivalent emission goals; a basic obligation for the development of national emission inventories; and the promotion of technology exchange.

Arctic programs

It is known that air emissions of cadmium, lead and mercury can travel a significant distance on Arctic air currents and thus can be deposited onto Canadian soil and water. Monitoring of such depositions and their effects takes place under Canada's Northern Contaminants Program (NCP) as well as other programs. Results of Phase I of the NCP, which is lead by the Department of Indian Affairs and Northern Development, were recently published in the Canadian Arctic Contaminants Assessment Report (CACAR). Canada is also the chair of the Arctic Environmental Protection Strategy (AEPS) comprised of the eight circumpolar nations. The Arctic Monitoring and Assessment Program (AMAP) under the Strategy carries out work similar to that of NCP but on an international scale.

While Canada has some idea of the quantities of these metals traveling through the atmosphere, the contribution attributable to Asian sources is quite uncertain. This needs to be elucidated in order to develop a policy for engaging these non-UN ECE countries in a control program, possibly through the United Nations Environment Program (UNEP).

Organisation for Economic Co-operation and Development

Canada is a very active member in the chemicals program within the OECD. The program includes harmonizing chemical testing and hazard assessment procedures; harmonizing of classification and labelling; developing principles for Good Laboratory Practice; co-operating on the investigation of existing chemicals (high production volume chemicals); as well as sharing and exploring possible co-operative activities on risk management of chemicals.

Within the OECD program for Guidelines for the Testing of Chemicals, Environment Canada with its federal partners helped to organize an OECD workshop, hosted in Ottawa, on the aquatic toxicity testing methods of metals. Environment Canada has also participated in examining the scientific basis for harmonization of classification and labelling of metals and most recently has become a member of the OECD Endocrine working group, to investigate endpoints of endocrine disruption and to harmonize risk assessment methods of endocrine disruptors. We have also lent our support to many risk management initiatives for several types of substances, including metals.

NORTH AMERICAN INITIATIVES

North American Regional Action Plan (NARAP) on Mercury

Canada has taken a leadership role in developing a North American Regional Action Plan (NARAP) on Mercury through the North American Free Trade Agreement and its Agreement on Environmental Cooperation. Through this agreement, Canada will be working with the USA and Mexico to reduce the emission of mercury to air. NARAP's are designed to provide specific actions to reduce risks to human health and the environment while recognizing the differing economic, ecological and political circumstances of the signing countries. The NARAP on mercury has four major objectives:

1) It provides a framework and approach for implementation of reduction measures;
2) It establishes the ultimate goal of preventing or minimizing anthropogenic inputs of mercury to the environment;
3) It establishes a mechanism to identify areas requiring further technical or scientific background information; and
4) It expands Challenges presented under *The Great Lakes Binational Toxics Strategy* to a North American Scale.

The Canadian Challenges promote a 90% reduction in the release of mercury by 2000 as well as provision for the study of long-range transport of mercury and remediation of priority sites. An amended "Phase II" NARAP on mercury, containing additional specific actions, targets and time frames will be completed in June 1999 by the recently formed implementation task group.

Great Lakes Initiatives

A number of initiatives focus on the Great Lakes Region. Under The Great Lakes Water Quality Agreement (GLWQA), amended in 1987, Canada and the U.S. pledged to seek virtual elimination of the discharge of persistent toxic substances to the Great Lakes. To this end, the Agreement included provisions for development of Remedial Action Plans as well as Lakewide Management Plans. Other programs in support of GLWQA include the Lake Superior Binational Program signed in 1991 which was a program involving nine persistent, toxic substances including mercury. In 1994, the Canada-Ontario Agreement (COA) committed the parties to a number of program targets. Among these was to seek a 90% reduction in the use, generation or release of seven [Tier I] substances by the year 2000. This list of substances includes mercury and alkyl-lead. For Tier II substances (including cadmium and tributyl-tin), Canada and Ontario agreed to collaborate with, and provide support for, voluntary programs by industry and others to reduce the use, release or generation of Tier II substances, and establish specific timelines and targets for achieving their virtual elimination. Most recent is the Great Lakes Binational Toxics Strategy (GLBTS) signed in 1997. This gives some focus to the long-range transport of contaminants into the Great Lakes region and provides some "out-of-basin" support to the Great Lakes programs. GLBTS contains a goal of reduction to natural levels through life-cycle management for naturally-occurring substances. As previously mentioned in association with the NARAP on mercury, GLBTS contains several challenges specific to mercury as well as several organic contaminants. It also challenges Canada and the U.S. to promote pollution prevention and the sound management of Level II substances (including cadmium and its compounds and tributyl tin).

NATIONAL INITATIVES

CEPA Priority Substances

One program within the federal government that assesses the human health and environmental risk posed by substances released to the Canadian environment is the Priority Substances Assessment Program under the Canadian Environmental Protection Act (CEPA). The Priority Substances List, mandated under Part II of CEPA, contains those substances which the Minister of the Environment and the Minister of Health feel that priority should be given to determine whether may be harmful to the environment or constitute a danger to human health. The list can contain chemicals, groups of chemicals, effluents, and wastes.

The purpose of the risk assessments is to determine whether the substances are "toxic" as defined under Section 11 of the Act which states:

"...a substance is toxic if it is entering or may enter the environment in a quantity or concentration or under conditions:

a) having or that may have an immediate or long-term effect on the environment;
b) constituting or that may constitute a danger to the environment on which human life depends; or
c) constituting or that may constitute a danger in Canada to human life or health."

The assessment process includes critically reviewing data on releases of the substance to the Canadian environment, fate of the substance in the Canadian environment, concentrations in the environment and toxicity of the substance to human and non-human endpoints. The process can include testing and additional research where it is felt that additional data are necessary to conclude whether the substance is "toxic".

If a substance is found to be "toxic" the Ministers may recommend that management strategies be developed to prevent or reduce the potential for harm to the environment or humans. Management strategies, which integrate socio-economic considerations, are developed in consultation with stakeholders and may include voluntary controls, process changes, substitutions, economic measures, regulations, guidelines, codes of practice, or a combination of these measures.

Twenty-five of the 44 substances on the first PSL were determined to be "toxic". Of interest here are the substances hexavalent chromium compounds, inorganic arsenic compounds, inorganic cadmium compounds, inorganic fluorides, and oxidic, sulphidic and soluble, inorganic nickel compounds. Options for managing the risk posed by these substances as well as other toxic substances were addressed in several different Issue Tables, including those established for the Base Metals Smelting Sector, the Electric Power Generation (Fossil Fuel) Sector and the Steel Manufacturing Sector.

The second Priority Substances List (PSL2) was published in December 1995 and contains 25 substances. Two of those substances of particular interest to the metals and mining sector are releases from primary and secondary copper smelters and refineries and releases from primary and secondary zinc smelters and refineries. In order to conduct the environmental risk assessments of these substances, a meeting was held in September 1997 with technical experts from industry, academia, government and non-governmental organizations to obtain their input on the scope and direction of the assessments. The leads for the assessments within Health Canada and Environment Canada also visited a facility in Sudbury, Ontario to gain a better understanding of the industrial processes involved and to determine whether monitoring data are available that would be useful in conducting the risk assessments. The environmental risk assessments should be completed early in 1999. We look forward to continuing support from our federal partners and industry in helping us complete these complicated assessments.

SUMMARY

The above initiatives, although not a comprehensive list, demonstrate that Canada is actively involved in several policy, assessment, and regulatory activities to conserve and protect the Canadian environment to help us achieve the goal of making sustainable development a reality in Canada.

International health and environmental regulations for metals: New challenges for copper.

Ricardo Badilla-Ohlbaum, Ph.D.
Executive Director
Chilean Mining & Metallurgy Research Centre (CIMM)
Avda. Parque A.Rabat 6500, Vitacura , Santiago, Chile
Email: rbadilla@cimm.cl

ABSTRACT

Copper has been used by humanity for a very long time and unlike certain other metals, it is an essential element for life. Its toxicity arises both from a deficit and excess copper exposure/intake. The focus of the current discussions in countries of the European Union and in some USA States, is based on the concerns, independent of the scientific facts and the economic status of these claims, of the negative influence of copper and other metals on health and the environment. These discussions have influenced the new regulations of copper in drinking water, and the discussions on the appropriate levels of copper in surface water, soils and sediments. To illustrate the challenges ahead for copper, the evidence by which the current USEPA, WHO and EU recommendation and regulations regarding copper in drinking water have been derived is presented and discussed. The gaps between the demands for new standards of health and the environment and the practical scientific tools to achieve these goals are presented, and the need to develop new regulating paradigms is also illustrated.

The impact of the possible restrictions on the use of copper in the EU because of health and environmental concerns in terms of the world market for copper is discussed. From the present use of copper, a 4.9% decrease in the demand in the world markets of copper in the near future can be estimated. From observed elasticity price/demand values, a 20-25% drop in the copper price will be followed by a return to normal prices in a period depending upon the rate of growth of the demand in emerging economies in which the consumption is low in comparison with developed countries. The importance of copper and the effects of this possible negative impact require renewed efforts by all interested parties to strengthen current research and to establish new regulatory paradigms based on up-to-date science.

Proceedings of
Metals and the Environment: An International Symposium
Edited by Jim Skeaff
The Metallurgical Society of CIM
100th CIM Annual General Meeting, 1998

RESUME

Le cuivre à été utilisé par les êtres humains depuis très longtemps. Le cuivre, au contraire des autres métaux lourds, est essentiel à la vie. Un certain degré de toxicité apparâit lorsque il y a un déficit ou un excès à l'ingestion de cuivre ou à l'exposition au cuivre. Des discussions, à l'heure actuelle, sont menées dans les pays de la Communauté Européenne et dans quelques états des Etats Unis, par rapport à l'influence négative du cuivre et d'autres métaux sur la santé et l'environnement. Ces discussions sont menées sans prendre en compte des faits scientifiques sur le sujet, ni des conséquences économiques des restrictions qui puissent apparâitre comme necessaires. Ces discussions ont influencé des nouvelles régulations sur la teneur maximal en cuivre de l'eau potable, des quantités maximales de cet élément dans la surface des eaux, dans le sol et dans les sédiments. Pour illustrer les défis auxquels le cuivre va être confronté, dans cet article on présente et l'on fait une critique de la façon dont l'USEPA, l'OMS et la Communauté Européenne, sont en train de réguler la teneur maximal en cuivre dans l'eau potable. On expose aussi la distance qu'il y a entre les demandes pour des nouvelles régulations de façon à proteger la santé et l'environnement et la capacité scientifique existante pour mèner à bien cette tache.

L'impact des possibles restrictions à l'utilisation du cuivre aux Communauté Européenne, dues à des régulations liées à la santé et l'environnement, considerées en termes du marché mondial du cuivre sont aussi exposées. Par rapport à l'utilisation que l'on fait aujourd'hui du cuivre, on peut estimer une baisse de la demande de un 4,9% dans le marché mondial dans le futur prochain. En utilisant le taux d'élasticité du prix à la demande, on peut s'attendre à une baisse du 20 à 25% du prix du cuivre. Avec le taux de croissance de la demande que l'on observe aujourd'hui, il faudrait 5 ou 6 ans pour récuperer Le prix d'équilibre de long terme va redevenir au niveaux actuelles suivant la croissance des economies en developpement, pour ces pays la consommation est bien plus baissse que pour les pays développés. L'mportance du cuivre et les éffets négatifs qui peuvent découler d'une politique de restrictions à l'utilisation de ce métal, font nécessaire redoubler les efforts des parties interessées, pour renfoncer la qualité et quantité de la recherche de façon à établir des nouvelles régulations qui soient reliées aux meilleurs bases scientifiques possibles.

HEALTH & ENVIRONMENTAL ISSUES RELATED TO COPPER

In a world characterised by increasing demands of new standards for the maintenance, protection and restoration of a high level quality of environment and health, there are a number of macro issues among opinion-formers throughout the developed world that have the potential to negatively affect world markets for copper. It is evident that in a number of issues of world-wide impact related to health, such as the discussions and actions taken by governments and agencies in relation to the bovine spongiform encephalopathy or "mad cow" disease in Europe, at present public opinion and policy makers perceptions play a much more significant role than science. This characteristic of the regulatory processes represents the biggest challenge to the mining industry and to the countries interested in supporting a sustainable development for mining production. It appears that for the first time since the recession of the 80´s, the copper industry is posed with a threat that can not be overcome only by internal adjustments to increase productivity or its competitiveness.

The main concerns are related to the negative influence of copper and other metals on the environment and health independent of the economic status and of pure scientific facts, perceptions of mining influence on uses of copper, and even questions regarding the sustainability of copper resource, use and growth. A summary of some important national and international environmental and health regulatory issues affecting copper is presented in Table 1.

Table 1 : Summary of Regional/National Concerns in relation to Copper, Health and the Environment

	National Issues	International Regulatory Issues
Germany	Drinking Water	
	Soils	Risk Assessment of Copper (EU)
	Sustainability	
		Drinking Water (WHO, EU) (Acute effects)
Sweden	Waste Water	
	Pollution Prevention Plans	Aquatic Availability
The Netherlands	Surface Water	Terrestrial Fate & Effects
	Soils and Sediments	
	Sustainability	Hazard Classification
USA	Pollution Prevention Plans (Individual States)	Basil Convention
		North Sea Conference

Copper has been in used by humans since the beginning of civilisation, and it is well known that it is an essential micronutrient for life on earth, including humans. Like most nutrients, its toxicity may arise from both a deficit and an excess of exposure. Although current knowledge and factual observations allow us to make science-based conclusions to establish and manage the risks of copper to the environment and heath, there are some uncertainties in the exposure levels at which some of its effects occur. These uncertainties and the difficulties in clarifying them in the short term, are the main reasons behind the difference in the perceptions on how to assess the risks of copper, and lead to the application of the precautionary principle to influence legislation and international regulatory issues, such as those shown in Table 1. In addition to the necessary review from time to time of the existing regulations, the aim of these discussions is to establish new levels for copper in drinking water and the environment (surface waters, soils and sediments), with the implication by many opinion formers that the existing regulations are not sufficient to protect health and the environment, based on exposure or effect data which is of questionable scientific value.

COPPER IN DRINKING WATER

A good example of the present trends in the revision of existing health and environmental regulations is the case of copper in drinking water.

The present debate has centred on the health effects and the levels necessary to protect consumers from the alleged negative effects of copper in drinking water. Copper found in drinking water (normally in levels from 0.05 to 1 mg/L) may arise from the leaching of copper-pipes and fittings depending upon the physicochemical characteristics of the raw water, the age of the system and the time of the day in which the water is taken from the tap (Lagos and Reveco, 1997), and raises the question of safety of copper as the prime material employed in water transportation systems in households worldwide.

For many years the worldwide practice, based on the absence of epidemiological data to relate any health effects of copper, was to include copper in the List of Substances and Parameters in drinking water that may give rise to complaints from consumers at a value of 1 mg/L (Table A2.5, WHO 1993), a level which was defined as a reference value to avoid staining of sanitary ware and complaints of the bitter taste of the water by consumers but with no health significance.

In 1991, the United States Environmental Protection Agency (USEPA) established a new health base Maximum Contaminant Level Goal (MCLG) for copper of 1.3 mg/L , a value which was derived from an old retrospective case study on a group of nurses who suffered acute gastrointestinal disturbances after consuming an alcoholic beverage contaminated by copper leached from the preparation vessel (Wyllie, 1954). Based on a replication of the original drink, exposure concentrations were derived and it was determined that the dose that caused the effect was 5.3 mg/L, from which the value of 1.3 mg/L is calculated using an uncertainty factor of 4. Although this work has been severely criticised by many epidemiologists (see for example Fitzgerald, 1995), this value was accepted as the basis to define the lowest observed adverse effect level (LOAEL) and to derive the value of the Lead and Copper rule by the USEPA and subsequently by many international expert panels of the WHO and other regulatory agencies. There are many other studies, also retrospective in nature, which show LOAEL values higher than 7.8 mg/L (Donohue, 1997), suggesting that the 1.3 mg/L level is possibly too low since under the provisions of the US Copper and Lead rule, water supply companies are obliged to adjust the characteristics of the raw water to comply with the level of copper delivered to the taps of consumers.

The questions regarding the possible chronic effects of copper in drinking water were first openly posed when a Revision of Drinking Water Quality Guidelines of the World Health Organisation was published in 1993 (WHO, 1993), in which copper was included - although provisionally - in the List of Health Significance parameters along with the most toxic inorganic substances known to man, such as lead, arsenic, cadmium, and cyanide. This provisional recommendation was adopted as a precaution against the possible role of high levels of copper (more than 8.5 mg/L in waters coming from wells and not from central distribution systems) in copper-associated childhood liver cirrhosis best described as Indian childhood cirrhosis (Pandit and Bhave.,1996), idiopathic copper toxicosis and German childhood cirrhosis (Müller-Hocker et al, 1987), which are rare diseases related to excess intake of copper through food and copper- contaminated drinking water. However, as established later, high levels of copper in drinking water is a necessary but not sufficient cause of the disease, as the ecogenetic character of these diseases involving a dysfunction of the homeostatic mechanisms regulating the absorption of copper was demonstrated recently (Müller et al , 1996).

As a result of this WHO publication regarding copper in 1993, the European Union followed that recommendation in its process of reviewing the current legislation on Drinking Water Standards (European Commission, 1994), and in the legislation reviewing process of the existing 80/778/EEC Directive in which copper has been included in the health significance parameters with the obligation to comply with an average value of 2 mg/L as a precautionary measure to avoid possible gastrointestinal effects from copper in drinking water. Unlike the USA lead and copper rule, in the case of the European legislation there is no obligation to adjust the quality of the raw waters in the areas of the EU with aggressive waters when needed, which may undermine the confidence in the use of copper in key markets of the European Union.

On the other hand, since the publication of the WHO Drinking Water Guideline, there have been many scientific discussions conducted internationally between 1993 and 1997, including a review by the International Program on Chemical Safety review process to establish an Environmental and Health Criteria on Copper (EHC) completed in Brisbaine, Australia in June 1996, at the request of the Chilean and French Governments. The conclusions of that process to be published soon (Becking, 1996) indicated that although there seems to be a narrow range between toxicity by deficiency or excess, there is a much larger risk for the population in the world of a deficiency of intake rather than an excessive exposure to copper. The same study concluded that there is a great uncertainty on the data to establish both the levels and effects of copper in humans, and in order to protect against the gastrointestinal effects of copper in drinking water, the upper limit of exposure is in the order of several milligrams per litre (meaning several more than 2-3 mg/L but not many).

Based on the IPCS EHC on Copper and the influence of the USEPA current views on copper in drinking water, the WHO Rolling Revision of the Drinking Water Quality Guidelines which took place in Geneva in April 1997 (WHO, 1997) maintained its provisional recommendation to classify copper in the List of Health significance parameters, this time to protect against possible gastrointestinal effects and ruling-out the possible direct role of copper in chronic liver damage. It recognises the lack of epidemiological data to establish the levels and effects of copper, and it mentions explicitly the prospective epidemiological studies currently being done in USA, Sweden and Chile, which upon publication, may shed some light into the issue of copper in drinking water, and could influence the new revision process of copper in drinking water which will take place in 1999.

In conclusion, copper is known to have emetic properties at relatively high concentrations, and the key issue in assessing health risk is to establish the exposure/intake from all possible sources (water and food). It is difficult to establish a clear cause-effect relationship for copper and to assess of health risks from the existing data, due to the lack of epidemiological evidence derived from prospective studies. The presence of micro-organisms and viruses in drinking water are the main cause of gastrointestinal health effects observed in drinking water (Payment, 1997), and the first two truly epidemiological prospective studies of the health effect of copper in drinking water carried out in Sweden (Pettersson, 1995) and Chile (Olivares et. al., 1998) have shown no correlation between gastrointestinal disorders and copper levels in the range 0 - 3 mg/L, which indicates that the levels of the current EU legislation aof 3 mg/L are quite safe, a factor confirmed by the absence of complaints by consumers of water of either gastrointestinal disturbances or taste problems.

THE NEED FOR NEW APPROACHES TO ASSESS ENVIRONMENTAL & HEALTH RISKS OF COPPER AND METALS

The principal paradigms to assess the risk of using metals, particularly essential elements like copper and zinc, in health and the environment, have been under scrutiny and revision during the last few years. This is due to an increasing trend to regulate metals by applying precautionary principles based on questionable scientific assumptions. One such assumption is that the effect of a metal on the environment depends on the total quantity of metal present rather than on how much of this metal is available to interact with living organisms. There is also a close relationship between health regulation (i.e. copper levels in drinking water) and environmental regulations (i.e. copper level in surface or run-off waters), since some opinion formers are interested in a low level of copper in the former as a means of decreasing the copper content in the discharge of sewage treatment plants into rivers or the sea.

Traditional methods for assessing health-based risk used for non-essential elements and compounds starts by defining an intake/exposure level at which no observed adverse effects (NOAEL) of biological significance are found or the lowest level at which adverse effects are observed (LOAEL). In addition, uncertainty (UF) or modifying factors (MF) are used to adjust NOAEL or LOAEL and define reference doses (RfD) at which the chronic intake/exposure levels considered safe or of no significant health risk. Uncertainty factors are used to adjust RfD data derived from experimental data of animal toxicity studies, or when data from humans are insufficient to account fully the variability in population, and therefore the usual values result in a 10-fold reduction of the acceptable exposure level for each effect that is being considered (USEPA, 1993).

In the case of copper or other essential micronutrients, low uncertainty factors are used since there are significant health risks associated with low intakes, and the RfD level is defined that not only prevents toxicity but to which the level of the deficiency in the population is minimised. This is called the recommended daily allowance or recommended dietary allowance (RDA or RDI), which is normally determined to satisfy the needs of 95% of the healthy population. Special considerations of upper or lower cutoff points should be given in defining the adequate range of intake (AROI) for physiologically active populations in infancy, pregnancy, lactation, ageing and exercise when pertinent, since each of these groups has special requirements. Thus, the AROI is the exposure/intake which in practical terms deals with the recommended level that will simultaneously prevent toxicity and deficiency.

AROI levels defined at the international level are not intended to address disease conditions or genetic alterations of copper metabolism which determine special sensitivity to copper deficiency or toxicity, such as Menkes syndrome or Wilson´s disease. In fact, Wilson´s disease is associated with toxic manifestations (Scheinberg and Sternlieb, 1996) at the level of exposure of normal populations as shown in Figures 1 and 2, so the occurrence of this disease can not be protected from copper exposure and needs a medical treatment to decrease the copper intake by patients.

It is known that the main route for copper intake is food, which in itself does not represent any problems since there have been no reports of copper toxicity arising from food intake. As shown in Figures 1 and 2, the AROI ranges are much narrower in the case of infants than in adults (with different values for preterm and term infants), which quite clearly illustrates the difficulties in establishing guideline values when the toxicity and deficiency and oral intake/exposure from usual dietary intake are taken into account (Olivares et al, 1997). The distribution of the exposure curves is estimated from the copper levels normally found in drinking water (from 0.1 to 1 mg/L) and the normal intake from food. Modern copper balance studies in infants and adults suggest that copper balance is affected by the type of feeding and the amount of copper supplied, and that the normal homeostatic mechanism regulates absorption, protecting the organism from copper deficiency and copper toxicity (Ehrenkranz et al, 1989, Turnlund, 1989).

The most encountered clinical manifestations of copper deficiency are anaemia, neutropenia and bone abnormalities (Danks, 1988), and in order to decrease the risk of these undesirable effects of copper deficiency, copper sulphate is added to preterm and child formulas as a common practise all over the world. In the case of copper toxicity, curves will be displaced to the right in Figures 1 and 2 (higher exposure values) in the series: taste threshold < nausea < vomiting < abdominal cramps < diarrhoea. The exposure ranges for these acute and reversible gastrointestinal effects are difficult to obtain precisely due to the intrinsic subjectivity of the effects among people and the lack of epidemiological data as discussed before. Chronic liver toxicity curves are not known since there are no valid animal models and no epidemiological evidence to draw such curves, but if there are effects, such values will be much higher that the exposure/intake levels shown in Figures 1 and 2.

The current regulation paradigm based on the LOAEL, UF and RfD does not necessarily take into account the complexities of essential micronutrient elements such as copper. The safe range of oral intake/exposure (AROI) for both infants and adults is shown in Figure 1. As demonstrated in these data, nearly 20% of the population receives too little copper, while excess copper is not a problem based on the isolated retrospective cases reported from copper

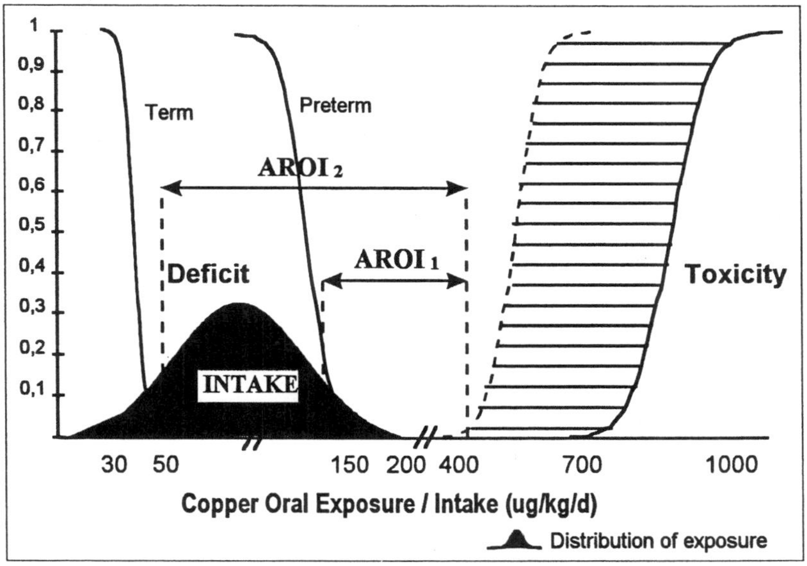

Figure 1. Safe range of oral intake/exposure (AORI) for term and preterm infants compared to usual dietary intake (gray hatched area following the deficiency/excess risk model). The uncertainties in our knowledge of toxic manifestation in humans is depicted by the striped area. Data on intake is from recent WHO/FAO/IAEA reports. As shown here, a significant proportion of the term infant population (20%), shown by the black hatched area, may be receiving too little copper, while excess copper is not a problem unless food or water is contaminated with copper. In the case of preterm infants most will develop copper deficiency unless given formula supplemented with extra copper (reprinted by permission from Olivares et.al.1997).

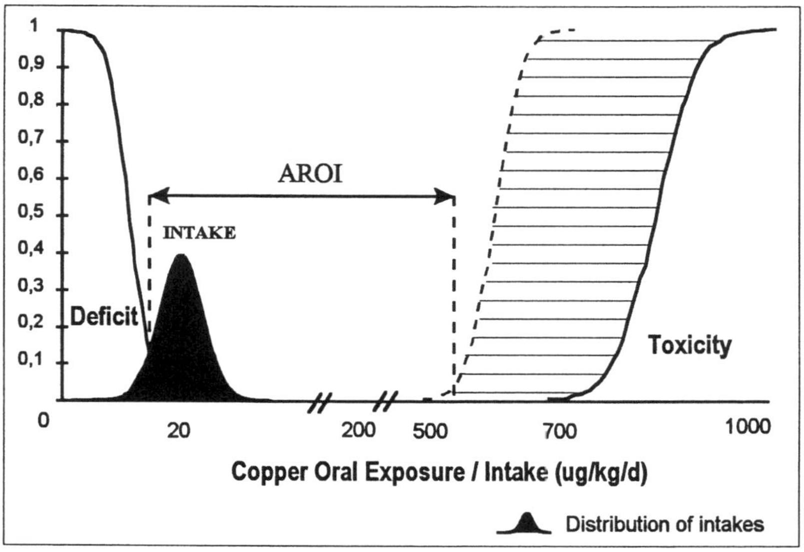

Figure 2. Safe range of oral intake/exposure (AORI) for adults compared to usual dietary intake (grey hatched area following the deficiency/excess risk model). The uncertainties in our knowledge of toxic manifestation in humans is depicted by the striped area. Data on intake comes from recent WHO/FAO/-IAEA studies. As shown here, a significant proportion of the population (20%), shown by the black hatched area may be receiving too little copper, while excess copper is not a problem unless food or water is contaminated with copper (reprinted by permission from Olivares et.al.1997).

in drinking water with millions of people exposed to copper in food and drinking water. The AROI for infants compared with the usual dietary intake shown in Figure 2 reveals a similar problem, and a significant proportion of the infant population is at risk from copper deficiency, which is partially avoided by adding copper to the formulas.

In conclusion, in the current regulating paradigms used to establish the NOAEL and LOAEL, the uncertainty factors to determine single levels of maximum intake are not the best way to determine the best exposure/intake range which will prevent health effects from deficiency and excess for essential elements such copper. In addition to continuing the research efforts to close the gaps in the existing knowledge on the health effects of copper, there is a definite need to develop new and more sophisticated paradigms for recommendations based on total exposure models and dietary requirements to establish the safe exposure/intake ranges of this element for a better quality of health in the population world-wide. This is also an urgent need for the case of environmental regulations of copper in soils, sediment and surface and marine waters, which are not being discussed in the present work for the lack of space. However, similar or even larger problems than those of the health regulations are experienced in this area.

THE POSSIBLE IMPACT OF THE NEW REGULATIONS

The use of wire, tubes, rods, profiles and rolled materials from copper and copper alloys in world-wide markets is distributed among construction, electric, electronics and other user sectors as shown in Table 2 (CRU, 1996). Recycling constitutes an important component of the raw material supplies of copper to the refining and manufacturing facilities, and accounts for 10 to 35 % of the use of copper, depending upon the price of copper which in turn largely defines the commercial viability of the recycling business .

Table 2. World Demand for Copper by use sectors

Use Sectors	% Distribution
Construction	35.8
Electric & Electronics	27.0
Industry- Machinery	16.2
Transport	11.3
Final Consumer Goods	9.8

The distribution of the use of copper shown in Table 2, indicates that one of the main sectors in which copper is used is construction, in which the main threats for copper consumption are also present. If a 75% restriction is imposed in the use of copper in drinking water transportation systems as a result of new regulations for copper in drinking water in the EU, a 50% reduction in the use of copper in roofs, gutters and drain pipes, and a 10% reduction in the usage of copper in car manufacturing and transportation as a result of the concern of copper in surface waters, this will result in a 7.5% reduction of the use of copper in developed countries and a 4.9% reduction in the world-wide usage of copper, as shown in Table 3.

Table 3. Estimated possible impact of the new health and environmental regulations in the European Union on and reduction of copper demand in world-wide markets.

Usage Sector	% of Total consumption at present in each sector	% Possible reduction in the demand for copper in EU in each sector.	% Reduction demand for copper in the EU.	% Reduction of total world-wide demand for copper
Construction				
Tubes and faucets	7.6%	75%	5.7%	3.7%
Sheets, rolls and profiles	1.3%	50%	0.7%	0.4%
Transportation	11.3%	10%	1.1%	0.7%
Total			7.5%	4.9%

It is difficult to estimate precisely the impact of a reduction in the prices of copper as a result of a decrease in the demand since the price of copper does not only represent the relation between the demand and the offer, due to the large influence in the price of the transactions in financial markets which normally accounts for more than 100 times the physical trade of the metal in world markets of copper. However, a 4.9% reduction in the demand for copper such as the one predicted in Table 3 in world-wide markets will significantly reduce the price of copper due to the inherent elasticity of the price with demand in the short term, which in practical terms varies from 0.20 to 0.25 (Vial, 1989; Codelco,1997). This reduction in copper demand could result in an initial 20-25% reduction in the world price for copper in the first 6-12 months. As a result of the normal increase in the demand in the uses not affected by restrictions such as the use of copper in countries not affected by restrictions, then depending on the rate of growth in the demand particularly from the emerging economies in which the consumption of copper is low in comparison to that of developing countries, the price will gradually increase until a return to normal is attained.

Since current copper mine production is about 12.5 million metric tons per year, the total current value of lost revenue of this cycle of low prices may be important, as caused at present by the economic recession in Asian Economies which significantly contributed to the drop in price in the last months of 1997. Lower prices of copper will not only affect the revenues of the copper producers but also the income of the economies of the southern countries, which contribute today with more than 55% of the total production, and depend on mining to satisfy the basic needs of their populations.

Copper is not only an essential element for life on earth but it is an important material to enhance the quality of living conditions for the population world-wide. The development of a sustainable copper industry requires a deep commitment by all concerned parties to establish the appropriate regulatory principles and practical scientific tools based on up-to-date science to establish and manage the risks associated with copper to health and the environment, in a framework of international collaboration.

ACKNOWLEDGEMENTS

The opinions expressed in this article are those of the author and do not necessarily represent those of the Centro de Investigación Minera y Metalúrgica (CIMM). I wish to thank Mr Santiago Torres from Codelco, Dr. Jorge Quiroz and Mr Jorge Berghamer and his team from the Chilean Copper Commission for their contributions into the analysis of copper markets, and my gratitude to Drs Gustavo Lagos, Ricardo Uauy and Manuel Olivares for their valuable contributions through many years working together in the scientific challenges associated with the health and environmental regulations of metals.

REFERENCES

Becking G, 1996, "Copper: essentiality and toxicity", IPCS News, 10: 4-5.

Codelco, 1997, "Elasticity of price/demand for Copper", private communication.

CRU, 1996, "Long term copper demand to 2010", Vols 1 and 2.

Danks, D. M., 1988, "Copper deficiency in humans", Ann Rev Nutr, 8:235-257

Donohue, J., 1997, "New ideas after five years of the lead and copper rule", in Proc. International Workshop in Advances in Risk Assessment of copper in the environment" (Lagos, G.E. and Badilla-Ohbaum, R. Eds), Reñaca, Viña del Mar, Chile, May 7-9, pp 265-274.

Ehrenkranz R. A., Gettner P. A., Nelli C.M., Sherwonit E. A.,Williams, J.E., Ting B.T.G. and Janghorbani M., 1989, "Zinc and copper nutritional studies in very low birth weight infants; comparison of stable isotopic extrinsic tag and chemical balance methods", Pedriatr Res 26:298-307.

European Commission, 1994, COM(94) 612 Final, "Council Directive Concerning the Quality of Water Intended for human consumption".

Fitzgerald, D.J., 1995, "Copper Guideline Value for Drinking Water: Reviews in need of review?", Reg Toxicol Pharmacol, :21; 177-179.

Lagos. G.E. and Reveco, F., 1997, "Human exposure model for copper in drinking water " in Proc. International Workshop in Advances in Risk Assessment of copper in the environment" (Lagos, G.E. and Badilla-Ohbaum, R. Eds), Reñaca, Viña del Mar, Chile, May 7-9, pp 237-252.

Müller, T., Feichtinger H., Berger H., Müller, W., 1996, "Endemic Tyrolean cirrhosis: an ecogenetic disorder", Lancet 347:877-880.

Müller-Höcker, J.Weiss, M., Meyer U., Schrammel P., Wiebecke B., Belohradsky B.H. and Hübner, G.,1987, "Fatal copper storage disease of the liver in a German Infant resembling Induan childhood cirrhosis", Virch.Arch A Pathol Anat Histopathol, 411: 379-385.

Olivares, M, Pizarro, F., Speizky, H., Lönnerdal B. and Uauy, R., 1988, "Copper in infant nutrition: safety of WHO provisional guideline value for copper content in drinking water", J Pedriat Gastroent & Nutrit, in press

Pandit A. and Bhave, S.,1996, "Present interpretation of the role of copper in Indian childhood cirrhosis", Am. J Clin. Nutr., 63:830-835S.

Payment, P., Siemiatycki, J., Richardson L., Renaud G., Franco E. and Prévost. M., 1997, "A prospective epidemiological study of gastrointestinal health effects due to the consumption of drinking water"., Int.Journal of Environm. Health, 7(1): 5-31

Pettersson, 1995, "The toxicity of copper" in Proc. International Workshop on Internal Corrosion in Water Distribution Systems (Hedberg, T., Vik.E.A., Ferguson, J.F. Eds), Göterborg, Sweeden, May 22-24, pp 29-32.

Quiroz, 1997 , "Possible Impact of Environmental and Health regulations affecting copper", Report to CIMM, Santiago, Chile.

Turnlund, J.R., Keyes, W.R., Anderson, H.L. and Acord, L.L., 1989, "Copper adsorption and retention in young men at three levels of dietary copper by the use of the stable isotope ^{65}Cu", Am J Clin Nutr, 49:870-878.

Vial, J, 1989, "World-market for copper: a systematic analysis", Cieplan, Santiago de Chile.

WHO , 1993, "Guidelines for Drinking Water Quality", Second Edition, Geneva, pp 46 and 174.

WHO, 1997, "Rolling Revision of the WHO Guidelines for Drinking Water Quality".

Mining and the environment: The public concern

Brennain Lloyd
Northwatch at Box 282, North Bay, Ont. Canada P1B 8H2

ABSTRACT

There is an emerging sensibility in civil society with respect to public interest rights and private sector responsibilities, and the same public concern that has provided watch-dogs for the forest industry for decades is now extending to the mineral sector. As operators on largely public lands, mining companies are being challenged to respond to the environmental and social expectations of stakeholders - the Canadian public.

Proceedings of
Metals and the Environment: An International Symposium
Edited by Jim Skeaff
The Metallurgical Society of CIM
100th CIM Annual General Meeting, 1998

INTRODUCTION

Mining and metals are a part of daily life. The benefits of mining in our metal-dependant society are with us in almost every activity - transportation, television, medicine, cosmetics, the list might be endless.

But so too are the impacts of mining and metal processing, and being metal consumers does not relieve of us of the responsibility to examine mining and metal use. Rather the reverse - as metal consumers we are called upon to understand the full chain of impacts that are spawned by our metal consumption, to examine where those impacts can be minimized or avoided, to place our use of metals in the greater context of environmental health, and to demand full stewardship of metals and metal products.

Throughout mining regions - in Canada, northern Ontario chief among them - mining has left a dismal legacy of hazardous sites and contaminated ground and surface water. The U.S. Environmental Protection Agency reports that mining has polluted over 180,000 acres of lakes in the United States, and the Ministry of Northern Development and Mines has estimated that there are six thousand abandoned continued mine sites in Ontario alone. The Canadian mining industry generates one million tonnes of waste rock and 950,000 tonnes of tailings per day, totalling 650 million tonnes of waste per year.

Public monitoring and participation in mining regulation has historically been extremely limited, and the mining industry is accustomed to operating without public scrutiny and with full access to public lands, and full opportunity to exploit public resources. While initiatives over the last several years, such as the Whitehorse Mining Initiative, have brought government, industry, labour and First Nations and the environmental community together to develop common goals and strategies to make mining more environmentally responsible, implementation is slow and legislation in some jurisdictions, such as Ontario, is being weakened rather than strengthened.

THE MINING SEQUENCE

Mineral extraction moves on a continuum of impact, from initial exploration through invasive exploitation to throwaway consumer products. From beginning to end - from abandoned mine sites to overflowing landfills – metal extraction, production and consumption poses threats and challenges to the natural world and the human communities who depend upon both the metals they use and the environment in which they live.

Exploration
↓
Mine operation
↓
Mine Infrastructure
↓
Mine Wastes and Water
↓
Mills and Processing
↓
Tailings and Waste Rock
↓
Refineries and
↓
Manufacturing
↓
Metal Products Stewardship

In the earliest phases - prospecting and exploration – impacts may be less severe, but they are also less regulated, and generally lack any requirement for rehabilitation of the site. And while impacts may be less severe, the site disturbance can be extreme, with complete removal of the vegetation and so total loss of ecological function.

Public concerns, values and other land uses often conflict with these activities. Trap lines, ski trails, sensitive natural areas such as old growth forests or special wildlife habitat - all are vulnerable to the paramount "rights" of mineral exploration. And the dividing line between prospecting, exploration and operating mine is a faint one, with more advanced exploration including sinking of mine shafts and actual ore production, but still lacking the public scrutiny and approval process that even the most limited timbering operations on public land would require. The mining process requires massive removal of soil and rock to retrieve the valued ore - a tonne of copper produces 99 tonnes of waste rock; a typical wedding ring required 3 tonnes of ore to be processed. The result is huge amounts of waste rock, which often contain toxic heavy metals and acid-generating materials. The ore goes through a number of processes on its way from mine to market, the first of which is a primary separation of the desired mineral from the other constituents of the rock - a process which produces large quantities of contaminated water and huge amounts of tailings, which are the "left-overs" of the separation process. Perhaps the greatest problems created by the mining process is acid mine drainage. Waste rock and tailings are exposed to oxygen and water, and generate acidic water which then frees up heavy metals. This effluent contaminates surface and ground water and area streams, rivers and lakes.

A Case in Point: The Serpant River
There are over two hundred million tonnes of radioactive tailings in the Elliot Lake Basin enough to cover the entire Trans Canada highway, coast to coast, two metres deep. The tailings cover over four square miles and have contaminated ten lakes and the Serpent River water system, which the International Joint Commission identifies as the largest single point source of radium in the entire Great Lakes. The tailings are up to 99% of the weight of the original ore, and retain up to 85% of the ore's original radioactivity. The tailings contain over a dozen radioactive materials, all of which have negative effects on plants and animals. The radioactive materials are called the "decay products" of uranium. The materials decay at very different rates, but many of them decay by producing a different radioactive material, causing a nuclear chain of materials and effects. Each material has a different half-life, a different "daughter", and affects life forms differently. For example, Thorium 230 becomes Radium 226 which becomes Radon 222, which becomes Radon progeny. The tailings are also acid generating, a result of the pyrite in the tailings creating acid when exposed to oxygen. The tailings also include many toxic chemicals and heavy metals, such as acids, arsenic, ammonia, nitrates, lead, zinc and cadmium.

The last stage in the "mining sequence" is also extremely critical, in terms of environmental damages. Metals occurring in compounds such as sulphide or oxide minerals, like copper or nickel, are isolated in their metallic state by smelting, where the sulphur is turned into the gas sulphur dioxide, which mixes with precipitation to become acid rain. In recent years, the emission levels of metal smelters have been substantially reduced, but cost and profit still draw the line on the toxic side of zero discharge.

A PUBLIC ROLE

From a citizen's perspective, the door is open and at the same time closed to public involvement in monitoring and affecting both decisions and operations in the mineral sector. For those with electronic access, they can monitor government registries, and on occasion see permits posted before they are granted for water-taking, waste disposal, and mine permits. But access is limited by the quality and timeliness of the posting, by the citizens access to not only the registry but then the daytime follow-through and technical expertise that is most often required. In isolated cases, advisory committees are in place which provide the public a recognized role in ongoing monitoring of the operation. But for most members of the public living in mining communities, the social barriers to speaking out or questioning mine operations may be the greatest impediment to citizen action and subsequent response from government and industry.

WHAT NEEDS TO BE DONE...

Mining and mineral exploration needs to be brought under public control, and decisions about mineral development brought into the realm of public influence. Just as landfills, highway expansions and pulp mill emissions are the subject of public examination, mining activities should be regulated in an open and transparent fashion, with a high standard of care demanded for the lands and waters that mining activities effect. That means zero discharge of toxic effluent, no net loss of habitat, and rehabilitation of mine sites. That means the "right to mine" claimed by industry and granted by government for the last century must be reconsidered within a reasonable array of rights - rights to a healthy environment, rights of the natural environment, and rights to clean air, water and land.

... by government

- identify all abandoned and orphaned mine sites, determine responsibilities and require that plans be developed to remediate the sites

- prohibit all exploration activities in wilderness areas, areas of scientific and natural interest, parks and nature reserves

- dedicate resources to meeting the Whitehorse Mining Initiative goal of setting aside from industrial development those protected areas required to represent Canada's natural regions

- require baseline flora and fauna studies and an outline of planned rehabilitation measures, in advance of any development or exploration work beyond initial staking

- make public notice of all advanced exploration obligatory, and require a closure (decommissioning) plan to be filed for public review before approval can be given

- require on-site treatment of all waste-water from mines and of all tailings leachate before release into groundwater or adjacent water bodies; apply the zero-discharge standard

- ensure that decisions about mining and mine permits are made in an open and transparent manner, with opportunity for public comment and requirements for adequate public notice

... by industry

- provide information to the public about mining operations in an accessible and comprehensive manner

- establish monitoring committees for mining operations which include citizens, labour representatives, First Nations, and government agencies

- involve the monitoring committee in the pre-operation development of rehabilitation plans for the mine site, and for environmental monitoring programs for the mine or mill's operating period

- establish an industry fund to remediate orphaned sites, and to do shared research on mine site remediation and acid mine generation

- self-define as a metal sector rather than a mineral sector, and include full product stewardship and chain of custody for metals to achieve full use and maximum use of extracted minerals

... by the public

- become familiar with mining regulations and with the opportunities for public involvement in mine permitting and monitoring

- identify mine sites, operating or abandoned, in your area, and learn about how they are being monitored, what permits have been issued, and how citizens are involved in decision-making

- develop a citizen monitoring program for mine sites of concern; monitor for heavy metals, acidity, and toxic contaminants, and share findings with the public, the company, and the regulatory agency

For more information, contact Northwatch at Box 282, North Bay ON P1B 8H2, tel 705 497 0373 fax 705 476 7060 email: nwatch@onlink.net

Northwatch is a regional coalition of environmental and citizen groups and individual members in northeastern Ontario. Founded in January of 1988, Northwatch addresses regional priorities: energy use and conservation; forest health, land use, and wilderness protection; waste management and water quality issues. Northwatch provides a representative voice on environmental concerns in northeastern Ontario, and supports local citizen's groups addressing environmental concerns in their community.

Global environmental challenges: An ICME Perspective

Scott Houston
Executive Director, Communications and Public Policy
International Council on Metals and the Environment
294 Albert St., suite 506, Ottawa, Canada K1P 6E6

ABSTRACT

A number of international negotiations and discussions will or have the potential to directly impact on the mining and metal industry's ability to conduct business. The impacts can be as simple as increased costs of doing business to loss of access to the resource or to loss of a market. The members of the United Nations Economic Commission for Europe are negotiating a "Heavy Metals Protocol" under the Convention on Long Range Transboundary Air Pollution. The immediate focus is on limiting emissions of lead, cadmium and mercury but there are also aspects of the protocol which cover technology and products. The proposed Convention on Prior Informed Consent has provisions which may restrict market access. The primary focus is on pesticides but metals can also be included under the category of industrial chemicals. At the Organization for Economic Cooperation and Development, discussions are progressing on the harmonization of chemical classification systems for substances and alloys yet fundamental concerns relating to hazard and risk are unresolved. Other national and international protection measures will also be touched upon.

Proceedings of
Metals and the Environment: An International Symposium
Edited by Jim Skeaff
The Metallurgical Society of CIM
100th CIM Annual General Meeting, 1998

INTRODUCTION

Before explaining what ICME's mission and mandate is, let me first put some of the issues that the mining industry is dealing with in a broader context.

LATE 20TH CENTURY TRENDS

The world has changed dramatically during the second half of the twentieth century and the rate of change is accelerating with each successive decade. There are three trends that are worth noting for the purpose of today's presentation.

First, is the international institution building that has taken place since the second world war and which has been largely a reaction to the perceived excesses of nationalism. Although some of the institutions that I will mention have their roots in the first half of this century, I think you will agree that their flowering took place from 1945 onwards.

Perhaps the two most important and well known institutions are the United Nations and the GATT, now the World Trade Organization. In the case of the former, there have been a succession of expansions of activity, from peace keeping to the establishment of the Environmental Program, the Commission on Sustainable Development etc. In addition there have been numerous negotiations: e.g. "The Peaceful Uses of Outer Space" and "The Law of the Sea Convention". Test ban treaties and nuclear disarmament are not so esoteric in nature and we all have a vital interest in their success. The GATT/WTO is really a child of the Havana Charter for an International Trade Organization, which was negotiated in the 1946-48 period. The initial intent was to re-build the world trading system in the post-war period. Although the Charter never entered into force, the General Agreement on Tariffs and Trade did so provisionally on January 1, 1948. Both institutions reflect a common view that the actions of nation states must be constrained and that the common weal can be improved by channelling activities towards objectives that are collectively agreed. The United Nations Conference on Trade and Development and the World Bank are two other institutions that I could mention, or the UN Economic Commission for Europe or the Organization for Economic Cooperation and Development etc. These institutions constrain the action of nation states.

Secondly, there is the issue of "globalization"- the trend over the last two decades towards a true mobility of capital and a "liberalising" of laws, constitutions and regulations at the national level as states realize that they now have to compete for capital as never before. "Foreign direct investment (FDI) has become more important than trade in terms of delivering goods and services to foreign markets, and it is the principal mechanism for organizing production internationally" is a quote from a report to the Trade and Development Board of UNCTAD entitled "Current international arrangements governing foreign direct investment". This phenomenon has engendered its own reactions at the international level - the General Agreement on Trade and Services (GATS), Trade Related Intellectual Property (TRIPS), Trade Related Investment Measures (TRIMS), Multilateral Agreement on Investment (MAI) and Multilateral Investment Guarantee Agency (MIGA), the latter an insurance provision. This is the response at the macro level and there are even responses at the micro level where the World Bank has embarked upon discussions on the development of guidelines governing environmental and social responsibility for FDI in developing countries. FDI dwarfs the level of investment that may be channelled through the development banks and it is not surprising that the institutional structures which are being marginalized by this process, are trying to maintain some influence over it.

Third, is the communications revolution as we enter the world of the Internet and the information super-highway. Just as this revolution brings us benefits of rapid decision-making and dissemination of information, investment funds and brokerage firms that operate 24 hours a day, moving millions of dollars around the world in a micro-second, it also brings us the added scrutiny of many interest groups who also treat the world as their own but from a very different perspective. Environmental groups are very active on the information highway and in the case of Greenpeace, they have been at the forefront of taking advantage of this technological revolution.

To sum up, the three trends are:

- agencies, treaties and conventions are serving to restrict the ambit of states;
- the market is global and corporations are increasingly multi-jurisdictional in nature; and,
- communications is increasingly making the world an open and transparent place.

If industry is to be a player, it has to become organized and just as effective at advancing and protecting its interests as others. This brings me back to my theme of how ICME and its members are responding to these trends.

ICME PARTNERSHIPS

ICME is an environmental non-governmental organization funded by its corporate members. The mandate is to promote sound environmental and related health policies and practices to ensure the safe production, use, recycling and disposal of metals. The focus is at the international level where we address environmental and health issues of a generic nature.

The membership of ICME now comprises 30 companies, resident on every continent. Most recently, we have had Brush Wellman and Battle Mountain Gold of the USA and Alcan of Canada become members. ICME represents a partnership amongst the industry in dealing with a range of environmental issues.

ICME has a committee structure that is organized in a manner to allow it to deal with a range of issues. The Board and Executive Committee provides overall guidance and help in priority setting. The Environmental Stewardship Committee develops and promotes strategies to achieve sound environmental policies and practices in non-ferrous and metal operations; contributes to international discussions about proposed regulations and policies; and, monitors environmental policies, practices and techniques. The Safe Use of Metals Committee develops strategies to promote the continued and safe use, transportation, recycling and disposal of metals and their products. It also assesses the regulatory issues so that the industry can contribute effectively to international policy and program development. The Committee on Health and Environmental Sciences has responsibility for developing and implementing scientifically sound tools, practices and policies for the protection of human health and the environment, taking into account the unique characteristics of metals and metal compounds. The committee emphasizes the importance of monitoring and assessing key toxicological, ecotoxicological, epidemiological and exposure information in policy development. The Communications And Public Policy Committee is responsible for developing communications programs, including methods for maintaining effective contact networks with stakeholders; develops responses to social, economic and public policy issues that interface with environmental concerns; promotes proper decision-making procedures and criteria; keeps members informed on key developments; and, analyses current issues in the context of social values and political structures which impact on public policy affecting members.

ICME cooperates with the international commodity associations recognizing that ICME deals with generic issues and they deal with commodity specific issues. There is often a complementary and mutually beneficial interface between these two areas of activity. Another group with which we have been active is the National/Regional Mining and Metal Associations. The intent is to grow slowly but effectively to ensure that there are solid linkages with national associations. Many of the international issues that ICME deals with will sooner or later have a domestic face to them and it is important to have early collaboration at this level.

THE ISSUES

Trade and Environment

The European Union had proposed that GATT Article XX, which provides exceptions to GATT obligations, notably the Most Favoured Nation (MFN) provisions of Article I, be amended to provide a broader exception for Multilateral Environmental Agreements (MEAs) with trade provisions. The reality of this proposal is to make environmental objectives legitimate grounds for trade discrimination and provide a coercive mechanism to encourage accession to MEAs or adherence to their provisions. The Singapore Ministerial in December 1996 did not agree with the European proposal.

The view of ICME is that Article XX of the GATT does not need amending. Both the chapeau and subsections (b) and (g) already provide sufficient scope to meet legitimate environmental objectives while providing protection against trade discriminatory or trade protective practices. [(b) necessary to protect human, animal or plant life or health; (g) relating to the conservation of exhaustible natural resources if such measures are made effective in conjunction with restrictions on domestic production or consumption]. An amendment could legitimize the coercive powers of members of an MEA over non-members of the MEA who are also members of the WTO; and, it could override national priorities and national sovereignty.

There are already 18 MEAs with trade provisions but only three of these raise serious issues of compatibility with WTO rules. As well, a number of negotiations are presently underway where trade provisions are likely to come up for consideration for inclusion, e.g. the UNECE Long Range Transboundary Air Pollution Heavy Metals Protocol, the UNEP/FAO negotiations on Prior Informed Consent and no doubt the up-coming negotiations on Persistent Organic Pollutants. Therefore, to eliminate the inappropriate use of trade measures in MEAs, more coherence and rigour is required in their negotiation, development and implementation.

Suffice it to say that changes to the GATT can have grave implications for the industry's ability to move product to market and one of the key concerns of ICME members is continued market access.

Work and discussions at the level of the WTO and OECD continue on this issue and will bear careful watching.

The Basel Convention

The Basel Convention is an environmental agreement that came into force in 1992. Its objective is to reduce the generation of hazardous wastes and restrict the transboundary movements of hazardous wastes to protect countries (mainly developing) that may not have the capability or technology to manage them properly. The basic principle of the Convention is to manage transboundary movements of hazardous wastes via a Prior Informed Consent (PIC) mechanism. This provision implies that the exporting country must not allow the export if it has a reason to believe that the waste will not be managed in an environmentally sound manner. The Convention further guarantees the sovereign right of each country to ban the import of hazardous waste into its territory. It also prohibits trade between parties and non-parties unless such trade is subject to bi- or multi-lateral agreement.

In September 1995, the Convention was further amended and in so doing established two additional requirements: i) an immediate prohibition of export of hazardous wastes for final disposal from OECD to non-OECD countries; and, ii) a ban on recyclable waste to be implemented on 1 January 1998. The amendment will come into legal force when ratified by three quarter of the member states (e.g. some 60 countries). However, this does not prevent countries from implementing the above decision on a unilateral basis.

Let me start by saying that no one disagrees with the basic objectives of Basel in regard to hazardous waste destined for disposal. However, the Basel Convention addresses more than that and I want to quote from a report by M.J.P. Guevara and M. Hart of the Centre For Trade Policy & Law entitled "Trade Policy Implications of the Basel Convention Export Ban on Recyclables from Developed to Developing Countries". They characterize the Basel Convention as "a badly-flawed Convention whose problems are compounded by the proposed export ban" on recyclables; and their principal recommendation to governments is that the "recommended course of action is refuse ratification of the amendment and to re-think the Convention".

In fact a number of governments, although not so blunt as the first statement, did make it known that with regard to the ban amendment, they would not consider ratification until the Technical Working Group completed its job in determining what waste materials would be subject to the ban. The work of this committee is proceeding, by all accounts somewhat satisfactorily, but its recommendations must still be approved by the Conference of Parties which can be a very political process. (COP 4 meets in February 1998)

ICME helps coordinate the input of the metals industry on this issue. Progress has been made at the various technical meetings. Importantly, the TWG has reaffirmed earlier decisions to generally exclude "non-hazardous" metal scrap and alloy recyclables from the scope of the ban decision and this includes precious metals, copper, iron and steel, nickel, aluminum, zinc, tin, magnesium and molybdenum, as well as zinc drosses and "wastes" from mining operations. I could expand upon this list to give you more detail on the specific nature and characterization of materials so far excluded but at least you can see that there is a growing recognition of the legitimacy of the commodity-like nature of metal scrap materials.

The Basel Convention makes a distinction between OECD countries and developing countries where the obligation to not trade is from OECD to non-OECD. Recycling is one of the key "R's" in any environmental policy of reduce, re-use, recovery and recycle. Yet the Convention decision-makers have ignored the fact that recycling plants are production processes which operate on an economy-of-scale basis and the catch-basin for raw materials often exceeds national boundaries. For recyclables, the ban provision is a serious disincentive. This is hardly the positive environmental outcome that most would want. As well, the scrap materials that are recycled supply part of the basic industrial infrastructure needs of many countries, whether they are classified as OECD or non-OECD members.

ICME has asked international experts to review the trade implications of the export ban and the applicability of article 11 bilateral agreements in the context of that export ban. Two studies have been published which serve to:

I) demonstrate the inconsistencies between the ban agreement and the WTO/GATT basic principles–information which could influence country decisions on whether or not to ratify the ban amendment; and,
II) identify, from a legal point of view, the conditions under which bilateral agreements for trading banned recyclables from OECD to non-OECD countries, would remain available, should the ban enter into force.

The final chapter on Basel remains to be written. However, one thing is clear, today the international community and industry will have to bear the cost of trying to correct the short-comings and clarify the inadequacies of the Convention while at the same time trying to make it work in a sensible manner.

UNECE-Heavy Metals Protocol

Some of the flaws inherent in Basel were at risk of being repeated in the UNECE Heavy Metals Protocol (HMP), a legally binding instrument the negotiation of which has recently been concluded (February 1998) and whose signature is expected to take place at a meeting of UNECE Ministers to be held this June in Denmark. While certain elements of the Protocol still give rise to concern, the final negotiated text represents an improvement over earlier versions.

UNECE stands for United Nations Economic Commission for Europe, a body comprising 54 member states. The HMP is another in a series which have been concluded under the Convention on Long Range Transboundary Air Pollution. The initial focus of the Protocol is on the heavy metals cadmium, lead and mercury.

ICME's preference would have been for the HMP to focus entirely on the reduction of long range transboundary air emissions as set out in the initial convention with the target being on point sources. However, throughout the negotiating process, considerable pressure was brought to bear from certain member states to include various articles and provisions as well as annexes on mandatory and non-mandatory product control measures which, for all intents and purposes, would have encouraged the ban and phase-out of all non-essential uses of the metals covered by the Protocol, as well as certain production processes.

The implications of a protocol dictating the management or ban of products and processes are significant for international trade, sale of technology, material usage and competition.

From the outset, industry expressed its concern regarding the scientific basis of the Protocol and that decisions were being based not on risk assessment but hazard identification criteria. In particular, industry was adamantly opposed to the inclusion of product control measures in the HMP. Because little emphasis was being put on science, it was argued that such decisions should be subject to strict criteria. More specifically, prior to recommending product control measures it should be demonstrated that:

- a product in normal use contributes directly to the long range transport of a targeted pollutant and that associated product-related emissions pose an unreasonable risk to human health and the environment; and,
- based on rigourous risk assessment and socio-economic analysis of the product in question and available substitutes, there are no risk management options other than product bans to address the problem at hand.

To illustrate, a key motivating factor for some delegates in encouraging the phase out of various products were concerns over emissions during waste disposal at municipal solid waste incineration plants. However, such concern overlooks the fact that application of fabric filter technology to reduce dust emissions from incinerators provides over 99% control of cadmium and lead emissions. Coupling this technology with powdered activated carbon (PAC) to control POPs (Persistent Organic Pollutants), results in mercury emissions being controlled to levels in excess of 90%. This suggests that the risk management options providing the greatest benefits at the least cost to society are not product bans, but point source emission controls on municipal incinerators.

Regarding production processes, some delegates attempted to introduce a ban on the use of the mercury cell process for the production of chlorine and caustic soda. This ban was proposed even though these plants meet current environmental regulations in Europe and in spite of the fact that they are normally being replaced at the end of their useful life by newer membrane technology. Setting a precedent on PPMs is very dangerous for the metallurgical and mining industry.

The final text of the Protocol, accomodated some of industry's major concerns. More specifically:

I) Contentious proposals which would have encouraged product substitution/minimization and economic levies on the heavy metal content in products have been moved from the main body of the protocol to an annex on non-mandatory product management measures. In addition, appropriate qualifiers/conditional phrases have been added to the accompanying text with the end result that the final wording is considerably less problematic;

II) Contentious proposals to include references to cadmium and lead products have been removed from the annex on non-mandatory product management measures;

III) Strict criteria and procedures have been adopted for adding other metals to the protocol.

IV) Strict criteria and procedures have been adopted for adding other metal-containing products to both the annex on mandatory product control measures and the annex on non-mandatory product management measures;

V) Proposals to ban the mercury cell process in the chlor-alkali industry have been dropped; and,

VI) Proposals that would have encouraged measures to control and/or reduce emissions of other heavy metals have been deleted.

The Protocol does, however, contain references to a number of specific products. In order to complement a soon to be announced European Strategy to phase-out lead in gasoline, consensus was easily reached to include lead in gasoline in an annex on mandatory product control measures. However, limitations on the use of mercury in various types of batteries have also been included in this mandatory annex. In addition, those mercury products that are subject of regulatory or voluntary action in various countries have been listed in the annex on non-mandatory product management measures.

Inclusion of various products at this time raises a number of fundamental questions regarding the Protocol itself, particularly with respect to internal consistency and its scientific foundations. However, it was suggested by some negotiators that the inclusion of these products was the price to be paid in order to obtain criteria and procedures to govern the addition of other metals and other metal-containing products to the Protocol. The latter is not insignificant as it is known that a number of governments are already suggesting that copper, zinc and nickel be added to the Protocol sometime in the future.

Prior Informed Consent

Under the auspices of UNEP/FAO, an international negotiating committee of governments has been convened with a mandate to prepare an international legally binding instrument for the application of the prior informed consent (PIC) procedure for certain hazardous chemicals in international trade.

The objective of PIC, as stated in the London Guidelines and the Guidance to Governments document, is information exchange, to advise governments of decisions taken to ban or severely restrict the use of chemicals for health or environmental reasons, and to request importing countries to advise the acceptability of future imports of those chemicals.

ICME has no difficulty in accepting the development of a legally binding instrument for PIC with the same scope, parameters and criteria as the voluntary scheme. Not surprisingly, however, once you open the door to negotiations, a number of new ideas come marching through!

The draft convention suddenly had an objective that was expanded to include the collective banning and phasing out of chemicals when governments were so agreed. This was fortunately struck from the draft as being beyond the negotiating mandate given to the committee by UNEP and FAO.

There are three categories of chemicals which are the subject of PIC - pesticides, industrial chemicals and consumer chemicals. Metals can be subject to PIC under the industrial chemical category. Under the voluntary program certain mercury pesticide compounds were already subject to PIC as was the mineral crocidolite under the industrial category.

Having learned the lessons of Basel, industry and quite a number of governments are taking a strong approach to the need for clear definitions, transparent and clear procedures and criteria, which are science based and include risk assessment. This is particularly important with regard to the question of how you nominate chemicals for inclusion in the PIC procedure. We are also concerned about a balance of responsibility between exporting and importing countries, irrespective of their development status.

Trade with non-parties is also an issue and again some governments seem to be willing to repeat the mistakes of Basel by allowing for trade discriminatory practices. On the other hand, there is also a proposal for clear language that would embody the WTO language on non-discrimination, artificial trade barriers, disguised protection etc. There is an intent to have members of this agreement "apply this Convention consistently with the WTO agreement. Nothing in this Convention shall be construed as affecting the rights and obligations of any member of the WTO under the WTO agreement, including its interpretative notes, as amended from time to time."

ICME would agree with this language being included in the Trade Provisions article of the convention.

Building Alliances & Partnerships

I want to now turn to other subjects where ICME is working jointly with others on key issues of concern to the mining and metals industry.

1. Risk Assessment Workshop

ICME, in collaboration with and with the advice of the World Health Organization, International Labour Office and United Nations Environmental Program, through the auspices of the International Program of Chemical Safety, the Australian Department of Health and Family Services, and Health Canada and with the formal support of the Chilean national Commission on Environment organized an International Workshop on Risk Assessment of metals and their Inorganic Compounds in November of last year.

The objective of the workshop was to provide a clear understanding of the issues and limitations related to the risk assessment methodologies for metals, and to develop a common understanding of a way forward which explicitly takes into account the special characteristics of metals.

The workshop made a number of recommendations including that: IPCS and the OECD should create an ad hoc advisory group of government, industry, academic and NGO risk assessors and scientists to advise on the feasibility of establishing a group to coordinate research, testing strategies and methodological developments related to the risk assessment of metals and their compounds; and, the establishment of an interdisciplinary expert group for the development of an Environmental Health Criteria document for ecological risk assessment of essential elements.

This follow-up work is continuing.

2. Tailings Workshop

Recent well-publicized tailings dam failures have prompted calls from some governments, international agencies and NGOs for regulations governing tailings design and management. The mining industry needs to and has responded to this issue proactively to overcome misconceptions and misinformation about tailings, tailings dams and cyanide bearing effluents.

In May 1997 in Stockholm, ICME and UNEP, with the support of the Swedish International Development Cooperation Agency, hosted a workshop on proper design, construction and maintenance of tailings retention systems. The workshop brought together over 90 international experts, regulators and mine operators from 27 countries, mainly developing countries with an interest in mining.

The workshop pointed to the need for promotion of best practices, particularly among small and medium sized companies, the use of risk assessment as a basis for decisions in the siting, design, and operations of tailings disposal systems, improved communication of technical and risk information and on emergency preparedness with the various interested parties, including communities, and enhanced cooperation between operating companies and regulators to make better use of the capacity of both sectors.

ICME is now preparing case studies of best practices in support of the above outcomes.

3. World Bank Guidelines

Here we have two areas where ICME has collaborated closely with the World Bank. Firstly, with respect to the Bank's development of environmental guidelines for mining, ICME and its industry members had the opportunity to provide input and review the progress of these guidelines at various stages in their development.

More recently, the Bank has responded favourably to ICME's views on the safe use and production of metals and is considering (internally) the role of ICME in delivering advice and training as part of an overall World Bank approach on mining.

Another initiative worth mentioning is the International Conference on Development, Environment and Mining organized by the World Bank, in collaboration with ICME in June 1994. The conference brought together over 300 people from some 100 countries and provided a forum for stakeholders to share ideas, experiences, information and solutions all aimed at enhancing the contribution of the mineral industry to sustainable development.

CONCLUSION

Having set the scene internationally, described the ICME mission and structure, identified some of the key issues that are confronting the industry, I want to conclude by categorizing the major industry challenges as follows:

1. Demonstrate that metals can be or are being safely produced, used, recycled and disposed of.
2. Credibly demonstrate that mining and metal production is supportive of the environmental, social, and developmental objectives of society.
3. Increase the international coordination and contribution of the industry in raising its environmental performance and in influencing policy development, the criteria and procedures used in the process, notably at the international level.

The industry is clearly facing pressures related to its markets for metals and in relation to its operations - whether justified or not.

We are in an information age where there are no secrets and where the public is flooded by information on risks.

Industry must build its response by acting credibly to convince governments that it is resolving the environmental and health issues of concern to society; by using sound science, analysis and provision of data to make its case on issues; by developing alliances with other organizations, agencies, community groups, indigenous peoples, and labour; by promoting fair, accessible and transparent regulatory procedures and criteria in decision making, e.g. the use of risk assessment that acknowledges the special characteristics of metals; by demonstrating the importance of metals to society, by stressing their economic, social and technical importance, the fact that they are among the most abundant of natural resources and that they are renewable in the sense of their recyclability; and by communicating effectively and responding in a timely manner to government and public concerns.

I trust that you will agree with me when I say that the industry through ICME is doing just that!

Thank you for your kind attention.

The management of releases from the metal-producing industries in Canada: Voluntary and regulatory measures.

Justyna Laurie-Lean
V.P. Environmental Affairs
Mining Association of Canada
350 Sparks St., suite 1105
Ottawa K1R 7S8

ABSTRACT

The voluntary initiatives of the members of the Mining Association of Canada (MAC) began with the development of an Environmental Policy in 1989. Since then, the Policy has been revised and several other initiatives undertaken. In 1997, MAC published its third annual report on environmental progress. This series of reports documents members' progress and commitments to reducing releases from their operations. The process of developing and implementing voluntary initiatives has been a learning experience, and the initiatives will continue to evolve. The benefits to industry and to society, while difficult to quantify, appear to be substantial, while the dangers turned out to be much smaller than originally feared.

Proceedings of
Metals and the Environment: An International Symposium
Edited by Jim Skeaff
The Metallurgical Society of CIM
100th CIM Annual General Meeting, 1998

1. Voluntary initiatives of the Mining Association of Canada (MAC)

MAC's voluntary initiatives can be divided into three broad categories. The first, and oldest, is MAC members' and support for multi-stakeholder, science-based , cooperative approaches to problem resolution. The oldest example is MEND. Its success gave MAC members confidence in the value of cooperative approaches. Since the initiation of MEND, MAC has supported many others, such as the Whitehorse Mining Initiative, AQUAMIN, the Aquatic Effects Technology Evaluation program, and, most recently, the Metals in the Environment initiative.

Another type of voluntary initiatives has been MAC members' efforts to collectively improve their environmental management processes. The initiative begun with a MAC Environmental Policy in 1989, and has since expanded to a revised Policy, an Environmental Management Framework, and other initiatives in progress.

The third type of voluntary initiatives is focussed directly on improving environmental performance. MAC members have been engaged in improving energy efficiency and are participating in the Voluntary Climate Change Registry. As well, MAC members expressed their commitment to continuing best efforts to reduce releases into the environment through participation in the ARET Challenge program.

MAC's participation in ARET forms the substantial topic of this paper. The paper describes MAC's experience with the program, the policy issues which arose, the problems that have been resolved, and the challenges that lie ahead. Details on our progress in release reduction, and on the other initiatives outlined above, can be found in the MAC Environmental Progress Report, which can be accessed on MAC's web site at http://www.mining.ca.

2. MAC and ARET - decision to participate

MAC participated in ARET from the beginning of the program's development. However, it was only one of several industries and stakeholders at the table. With other stakeholders focussed on issues related to organic chemicals, MAC was often isolated when trying to raise issues peculiar to metals and metal compounds. MAC therefore faced a difficult decision on whether to participate in the resulting ARET challenge. The following were the concerns MAC felt at the time, some specific to our industry, and some shared by others:

a) Lack of speciation

There was not sufficient metal-species specific data available to allow more than very crude grouping of metal compounds in the ARET lists of target substances.

b) "Persistence" and "Bioaccumulation" criteria

MAC was concerned that participation in ARET would be held as acceptance by MAC of the application of these hazard criteria to metals and metal compounds.

MAC decided that its concern about these first two issues was real, but that it could still participate in ARET on the basis of its commitment to the reduction of releases to the environment, without endorsing the exact methodology for arriving at the lists of target substances.

c) Uncertainty whether the reduction target could be met

At the time, most MAC members were uncertain that they could meet the set target (50% reduction by 2000 from base year). Some were sure that they could not meet the target.

MAC decided to accept the challenge and commit to best effort. If best effort fell short of the target, we would at least have demonstrated progress.

d) Fear of bad publicity

MAC anticipated that, for certain substances, MAC members would be shown to be the largest sources of release, and feared that this would lead to bad publicity.

e) Fear of misunderstanding and misuse

The experience of U.S. industry with the Toxics Release Inventory raised fears that any published release data would be misunderstood and misused.

On these last two issues, MAC members decided to take the risk, and to face the results. The decision was finally made that MAC would encourage its members to participate in ARET.

Experience with the publication of the first ARET report, and subsequent participation in the program, demonstrated that many of the anticipated problems did not materialize. On the other hand, unanticipated problems did arise and continue to become evident, requiring ongoing effort.

3. The problems we anticipated

a) Lack of speciation

Inadequate speciation of metals and metal compounds is an ongoing problem. Work in the OECD on hazard criteria, and the Metals in the Environment Initiative, may uncover ways of addressing this issue.

b) "Persistence" and "Bioaccumulation" criteria

With the invaluable help of the Canadian government, and with the effort of the world metals industry, this issue was placed on the OECD agenda. Work within the OECD has led to international recognition that "persistence" and "bioaccumulation" are not useful hazard criteria for metals and metal compounds, and to work on alternative criteria.

c) Uncertainty whether the reduction target could be met

Not every facility or company could meet the reduction target for every substance. However, we found that the industry as a whole could meet the target. Moreover, once companies engaged their best efforts, they found ways to reduce further than they had originally thought possible. The release forecast for 2000 was revised with each iteration. Companies' confidence in what they could accomplish increased and their enthusiasm and commitment rose. In the Base Metal Smelter Strategic Options Process, the smelters committed to an 80% reduction in release of target substances by 2008 from the base year, and further reduction beyond 2008.

d) Fear of bad publicity

The constructive and positive response from governments, the media, and certain environmental groups to the first ARET report was very encouraging. Their focus was on the industry's transparency and progress. This positive response was even evident internationally. Whenever MAC's release numbers are presented in international fora, other countries appear greatly impressed by what we have achieved.

e) Fear of misunderstanding and misuse
The ARET program has been fiercely attacked, but not in the ways anticipated.

4. The unanticipated issues which arose as we gained experience

a) Difficulty of estimating releases

Several years of experience with reporting releases and peer group discussions have shown that our release data fall short on quality and consistency. The reasons are many:

- Our industry has very large and very variable material flows, and the target substances are released at very low concentrations. Achieving estimate accuracy is difficult, and the margin of error will always remain large.
- Some material flows in our industry, such as stack emissions, are difficult and expensive to sample.
- The concentration of some substances is near or below detection limit. In such cases, an approximation of the total quantity released can be made, but the result is not useful in tracking trends over time. This problem will increase as releases are reduced.
- Most industry sampling is in response to regulatory requirements. Such sampling is designed to check that the concentration of a substance is below the regulatory limit, and is not designed to facilitate the estimation of the total quantity released.
- Each company and facility may interpret requirements differently and may use a different sampling and estimation methodology.
- We do not have complete knowledge of all process streams, particularly of fugitive emissions.

As facilities examined their processes and release points, and explored reduction options, they gained a better understanding of their releases, and revised their estimates accordingly. MAC formed a Data Integrity Group to explore the issues of release reporting and to develop protocols that will give us greater confidence in the consistency and quality of our release data. Protocols for arriving at release data would be very helpful to developing a practical process of verification. Once the validity of a protocol is established, it is relatively easy to verify whether it has been used to estimate a release. Without protocols, verification would require an examination and subjective judgement of the adequacy of the methodology used for each release estimate, and would still leave the question of consistency unaddressed.

b) Hostile response of certain environmental groups

While some environmental groups welcomed industry's efforts and the resulting reductions in releases, the Canadian Environmental Network Toxics Caucus has been extremely critical of the ARET program. The Toxics Caucus sees ARET as lowering the pressure for regulatory action or as giving governments an excuse to avoid regulatory action. The Toxics Caucus believes that regulations would achieve larger reductions, would apply equally to all companies and thus not permit lower degrees of effort, and that regulations should focus on reducing the use of toxic substances, rather than on reducing their release.

The specific criticisms made of the ARET program need to be viewed in this context:

Criticism that the base year should have been different ARET picked a date earlier than the program starting date in order to be fair to companies which had just undertaken major reduction efforts as compared to companies that had done nothing previously. Not every company used the same starting date. At least in our sector, companies used later starting dates where the facility did not open until after 1988 or where data were not available for 1988.

It does not matter whether the base year was right or not. ARET is an ongoing program. Companies continue to make efforts to reduce their releases. MAC members have made specific commitments to 2008. Surely the argument over 1988 versus 1994 is irrelevant now.

Lack of verification The lack of verification is a valid criticism. However, some degree of verification does exist through NPRI and through provincial reporting. Moreover, it should be recognized that there is no incentive for companies to mis-report. If current releases are reported as greater than they really are (presumably to allow for bigger reductions to be demonstrated in the future), the company risks becoming the target of regulatory and public attention as the "worst". If, on the other hand, current releases are reported as lower than they really are (presumably to allow the company to claim bigger reductions achieved), then future reductions will appear smaller than they really are.

The issue of verification does not arise countries other than the U.S., because there are no inventories of releases, voluntary or otherwise. At best, the Europeans can produce so-called "inventories" created by multiplying aggregate production figures by emission factors.

Reductions caused by regulations and not voluntary measures This criticism is partly true. The release reductions reflect a combination of responses to regulatory changes, economic-driven modernization, as well as environmental protection efforts. It would be impossible to objectively allocate the reductions by cause. Again, if we compare our progress to other countries, it is obvious that either this voluntary initiative was successful, or Canada has outstandingly stringent regulations.

Level playing field It is true that no voluntary program will have 100% participation. However, that does not invalidate the voluntary program. Reduced releases will result in better environmental protection. Surely we should not stop trying just because not all companies are trying. In the case of MAC, all the large emitters are participating, and the participation of all companies in the sector would not have a material effect on the total releases.

Mixed monitoring and reporting methods We have recognized the variability in companies' measurement and estimation methods, and are taking steps to address this. Thus, had it not been for the ARET program, any inconsistencies may never have been identified.

It should also be noted that there will never be, and should never be, only one measurement and estimation protocol. Different approaches are essential, depending on whether the substance in question is in high or low concentration, in a small or large volume stream, and whether the concentration and volume vary or are consistent over time.

Lack of consequences In my opinion, the objective of voluntary initiatives is to change attitudes, so a "gotcha" approach does not seem to be appropriate. Our member companies look on their voluntary commitments seriously as requiring best efforts. They recognize that their ability to use a "due diligence" defence would be compromised if they evidently failed to adhere to their commitments.

Delay of regulatory measures In my experience, regulatory approaches are much slower, and achieve much less than what ARET already has done. However, it is impossible to prove "might-have-beens". One can only look at what has been accomplished by countries that tend to favour regulatory approaches, and in that comparison, the ARET program looks good.

5. Future Challenges

I see four key challenges that will have to be addressed in the next five years:

a) Verification

Some process of verifying release data will have to be developed and implemented. The key will be to make it credible but cost-effective. If the MAC Data Integrity Group is successful in developing protocols for estimating releases, the development of a credible and cost-effective verification process will be greatly simplified.

b) Continuing to meet the expectations of other stakeholders

It can be expected that diminishing returns will slow the rate of future release reduction. As releases are lowered, changes in production levels, and unexpected occurrences, will make progress less even. New breakthrough technology will enable further large reductions, but such technology needs a long lead time to develop and implement, so the large step reductions will be infrequent. Stakeholders, on the other hand, expect steady progress. While giving credit for past achievements, they will focus on what we did last year and on what we are doing this year and next year. We can therefore expect occasional misunderstandings and friction.

c) Dealing with "peripheral" issues

As an industry, we focus our attention and efforts on what we consider our highest environmental priorities. Other stakeholders, focussed on their priority issues, frequently ask us to work on issues which are a low priority for our industry. For example, governments are very concerned about smog in urban areas, and have asked all industries to address emissions of volatile organic compounds. At most sites in the mineral industry, smog is not a problem, and our emissions of volatile organic compounds have little impact, particularly in contrast to our other emissions.

We will have to deal with each of these requests with sensitivity, deciding which ones to incorporate in our work and which ones to ask to be excused from addressing.

d) New problems from new technologies

Some new technologies which help us address current priority releases may create new problems, such as releases of "persistent organic pollutants". We will have to become familiar with issues we have not had to deal with before, and work with other stakeholders to select the best path forward.

6. Conclusions

Debate will continue over the value of voluntary and regulatory approaches in environmental protection. MAC believes that both approaches have value when used appropriately. Which is best depends on the particular aspects of the problem being addressed.

Regulations are useful in setting a performance "floor". They are easiest to set when consistent requirements can be set for all facilities and when the benefits are identifiable. Regulations are difficult to set when each facility is very different, when costs of compliance are very variable, and when benefits are difficult to identify and measure.

Regulations are limited in the improvement they can effect. They are normally set to be achievable by the weakest company. In practice, governments are very reluctant to issue regulations which will result in the closure of an operation and result in job losses. Thus regulations provide no incentive for companies with more resources or options to go beyond the regulatory requirements.

Negotiating regulations forces each side to focus on proving what can or cannot be done at an absorbable cost with practical technology.

In contrast, voluntary initiatives can activate industry innovation from people who know their process best and can encourage each company and facility to pursue best effort. The result is better environmental progress at a lower overall cost to the economy.

Dealing with releases from metal industries: Sources of metals to the environment released during production and/or combustion

Possible effects of global climate change on mercury re-emission and cycling

J.O. Nriagu
Department of Environmental & Industrial Health, School of Public Health, University of Michigan, Ann Arbor, MI 48109 U.S.A.

ABSTRACT

This presentation will explore the relationships between mercury emissions, mercury cycling and global climate change (GCC). Among other things, GCC is expected to result in (1) an increase in the amount of UV-B radiation reaching the earth's surface, (2) a change (increase) in temperature of the atmosphere and earth's surface, and (3) a change in the hydrological regime, especially the incidence of catastrophic weather incidents such as floods, storms and droughts. GCC is expected to enhance the remobilization and bioaccumulation of mercury in ways that could increase the risk of exposure of many people (especially native Americans and anglers) to this pollutant.

Proceedings of
Metals and the Environment: An International Symposium
Edited by Jim Skeaff
The Metallurgical Society of CIM
100th CIM Annual General Meeting, 1998

(I) Current industrial emissions of mercury from industrial sources in North America (estimated to be 428 ton per year) are believed to be comparable to the flux from natural processes (sources) and GCC is expected to change this ratio. The reservoir of pollutant mercury in ecosystems of North America should sustain the increased rate of emission by natural processes for a very long time. Figure 1 shows the most intensive emission and deposition sites in the United States. GCC may change the current deposition pattern and can result in further dispersion of already deposited mercury.

(II) A number of sites in North America are heavily contaminated with mercury (including abandoned gold and silver mines and chlor-alkali plant graveyards) and a change in climate may lead to increased exhalation of mercury from these so-called "chemical time bombs". Figure 2 shows the locations of some of the more important gold and silver mines in South and Central America.

(III) Many natural and anthropogenic sources emit mercury in a form that can undergo photochemical reactions in the atmosphere. The removal of mercury from the atmosphere is driven by the formation of reactive Hg(II) species by direct and indirect photochemical processes which are temperature-dependent. GCC can thus alter the current deposition pattern for atmospheric mercury.

(IV) Global climate change is expected to trigger an increase in rates of biogenic production and release of volatile mercury compounds, especially the methylated compounds which are most readily absorbed by fish. Marked seasonal differences (summer increases) that have been reported in airborne mercury levels in a number of studies may be a harbinger of expected outcome of global warming.

(V) Increase remobilization of previously deposited mercury may convert the northeastern region of the United States from being a sink to a giant area source of mercury. The mercury so released will be dispersed more widely and may change the deposition rates in other areas of the continent, especially eastern Canada.

(VI) Bioaccumulation of mercury by fish is closed linked to production of dissolved organic carbon (DOC) and water temperature, and these habitat characteristics are strongly influenced by GCC.

(VII) Extensive flooding of coastal and low-lying areas would exert the most drastic influence on the mercury cycle by increasing the efficiency and rates of mercury methylation as well as the levels in water, zooplankton, benthic invertebrates and fish in the newly formed bodies of water. The downstream effects of the flooded areas may expose a significant fraction of the rich fisheries resource of Atlantic seaboard to risk of mercury contamination.

(VIII) In temperate lakes, changes in food chain structure and function tend to be non-linear so that small changes due to climate may result in rapid and drastic changes in mercury methylation/demethylation and bioaccumulation rates.

(IX) Tropical and arctic ecosystems are particularly sensitive to mercury pollution. Because of the unique features of the food web, top predators of tropical and arctic ecosystems are more vulnerable to mercury compared to temperate species, and global warming can further exacerbate the exposure of the most sensitive organisms to pollutant mercury.

Human activities have changed the natural biogeochemical cycle of mercury in many ecosystems. GCC is expected to enhance the remobilization and bioaccumulation of mercury in ways that could increase the risk of exposure of many people (especially native Americans and anglers) to this pollutant.

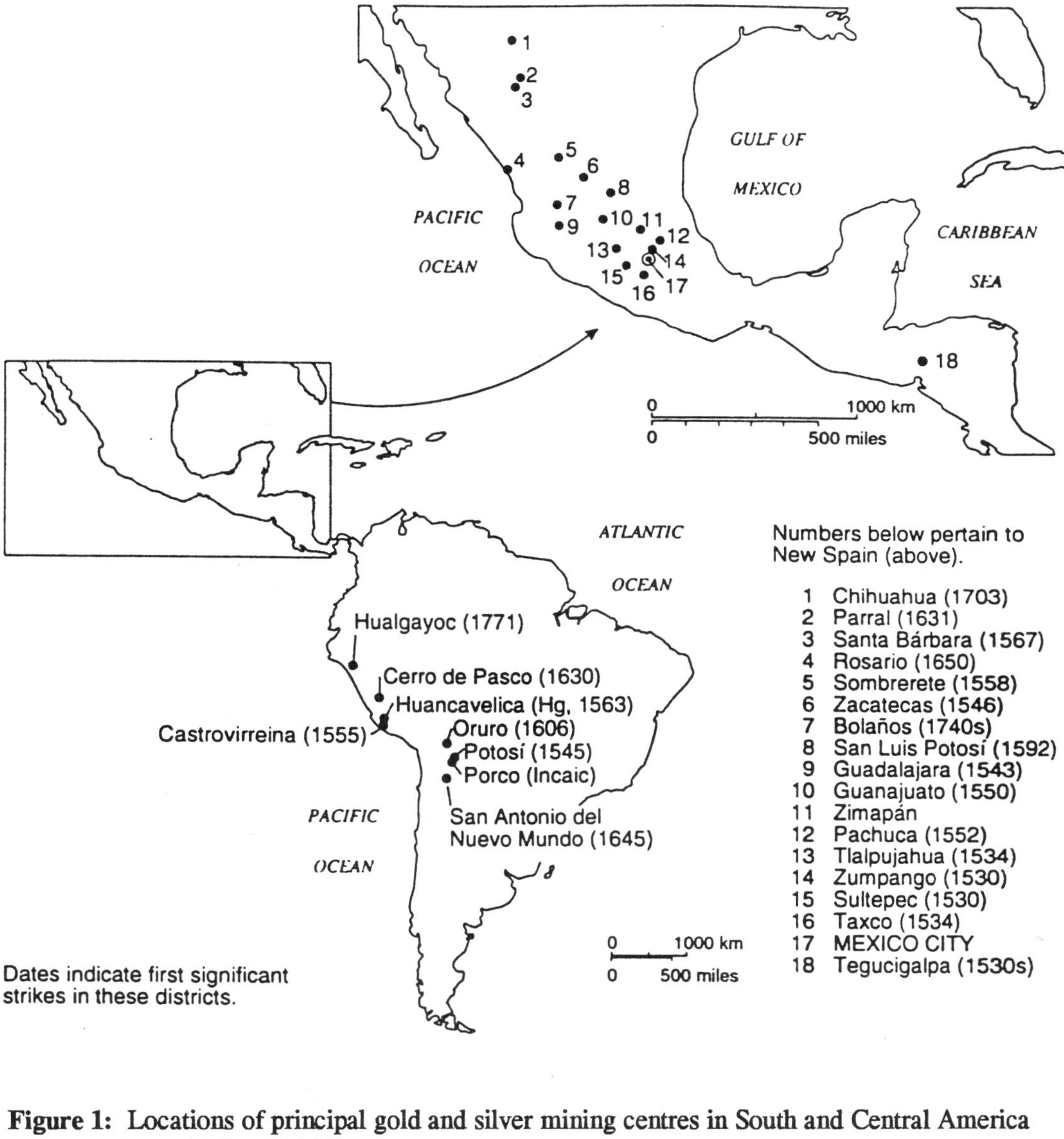

Figure 1: Locations of principal gold and silver mining centres in South and Central America (from Nriagu, 1994b).

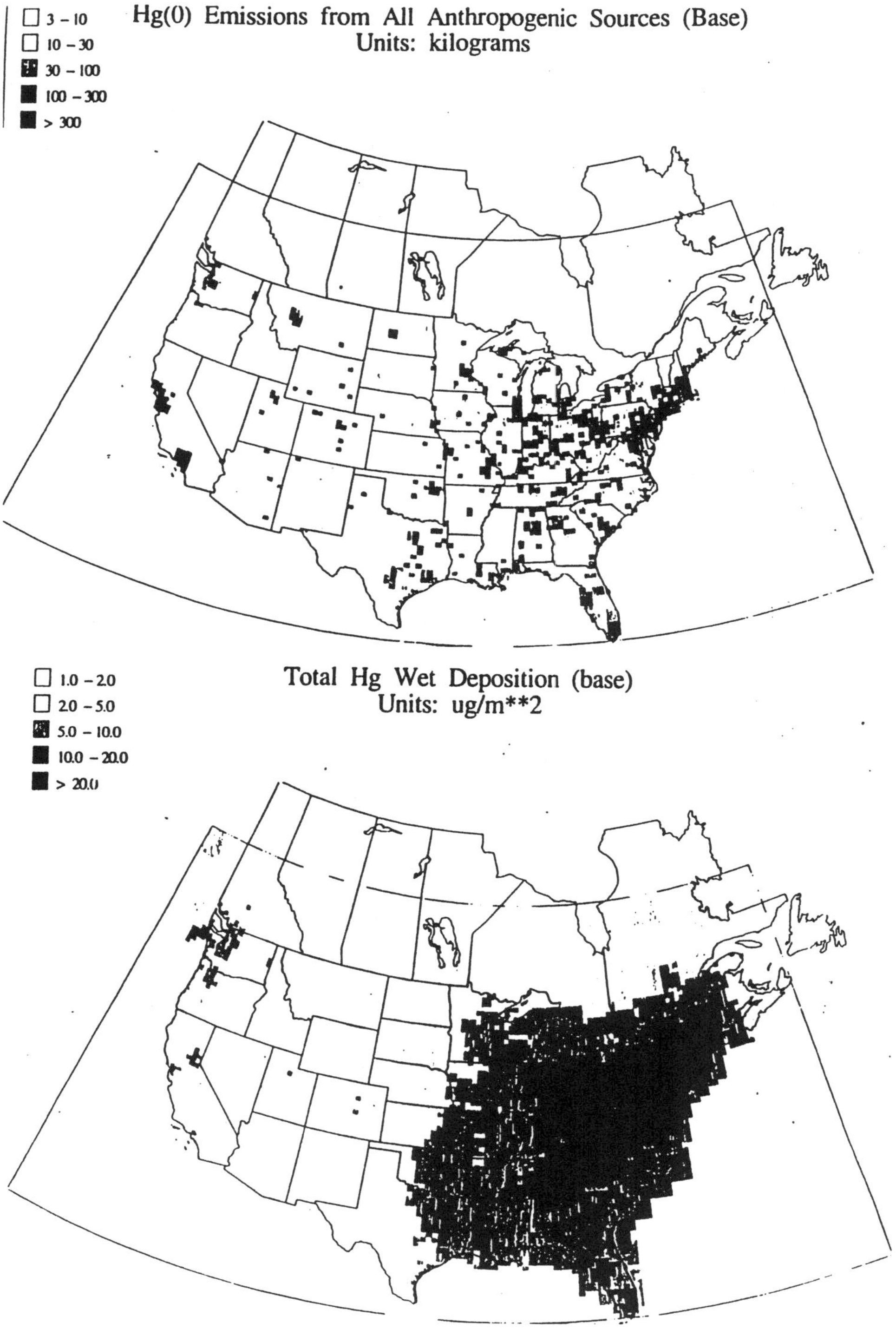

Figure 2: Emissions of elemental mercury from industrial sources (top), and calculated wet deposition of mercury (bottom) in the United States (from US EPA, 1994b).

Metals and municipal solid waste incineration

A.J. Chandler QEP
A.J. Chandler & Associates Ltd.
Willowdale, Ontario, Canada M2M 2H1

ABSTRACT

It has been suggested that municipal solid waste [MSW] management systems, and in particular incinerators, are a major source of trace metals in the environment. The reasoning is that, since these materials are found in MSW they must be released to the environment when the waste is burned. This argument appears to ignore the fact that MSW incinerators are amongst some of the most highly regulated air emission sources in modern society. These facilities have had to employ sophisticated emission control systems that reduce the concentration of heavy metal species and trace organics to levels that are now taxing the limits of measurement technology.

The presentation provides information on modern EFW emissions. The modern fabric filter baghouse reduces emissions of particulate matter to below 17 mg/Rm^3 @ 11% O_2 and that of heavy metal species to even lower levels. Efforts to improve organic emission control measures has prompted the use of sorbents such as powdered activated carbon [PAC] which sorbs mercury and reduces its emissions by in excess of 90%.

This presentation will also discuss the partitioning of metals in the incinerator, and the results of some trace metal spiking studies undertaken at the Burnaby incinerator in 1991.

Proceedings of
Metals and the Environment: An International Symposium
Edited by Jim Skeaff
The Metallurgical Society of CIM
100th CIM Annual General Meeting, 1998

INTRODUCTION

There is no doubt that the statement, "Metals are present in Municipal Solid Waste [MSW]" is true. The question is however not, "Are trace metals present in MSW?", it should be: "Are metal releases from MSW management alternatives a serious concern?" Furthermore, if such releases are a concern, the secondary question that should be addressed is: "How can we best reduce these emissions?" Some would suggest that we must immediately formulate new products that contain lower levels of trace metals. Others have gone as far as to suggest bans on certain metals in consumer and industrial products. Some in industry, such as the dry cell battery manufacturers, have responded positively to these concerns and removed mercury from their products. Such an orderly evolution of products is good for society.

Unfortunately, those who would apply bans to products are not willing to let the economy and industrial innovation come up with these solutions. They also are not able to say unequivocally that the replacement will reduce the risks to society and the environment. Furthermore, they cannot positively state that such bans will produce the desired results. Without a thorough knowledge of the source and fate of trace metals in the MSW management system, society could invest vast sums of money in removing minuscule parts of the problem. Indeed, it has been argued that fluorescent bulb separation programs and dry cell battery return programs might have little, if any, effect on mercury emissions from MSW incinerators[1]. Research undertaken in Canada in 1991 showed that more than half the mercury present in the municipal solid waste [MSW] stream entering an incineration facility was in natural materials[2].

Recent studies in Canada suggest that up to 40% of the residential waste stream is comprised of organic matter either yard and garden waste or food waste[3]. The WASTE Program[2] data indicates that the median for yard and food waste at Burnaby was about 20% but individual 150 kg samples contained between 1% and 54% of these materials. Wood waste and fines, the small-sized, not easily separated portion of the waste stream coming largely from the environment, brought the median of materials that are not man-made to approximately 37%. Man-made materials, including rocks, sand and dirt from construction waste contribute approximately 57% of the total waste stream.

Paper products comprise 61% of the man-made materials in the waste or about 38.5% of the total waste stream; plastics contribute 13% of the man-made total and the glass and metal combined another 7%. Overall, paper and packaging materials accounted for over 51% of the total waste sampled. The balance is made up of miscellaneous materials including textiles and rubber, household hazardous, small electrical and construction debris.

The WASTE Program[2] selected at least one sample from most components for analyses of both total metals concentrations and the environmental availability of trace metals using SCEP. Detailed data for arsenic, cadmium, mercury and lead concentrations in all components and size fractions, expressed as percentage of the total of that element in the waste stream, is provided in the main report and elsewhere[4]. The results for the major component groups are given in Table 1. These data show that organics contribute the bulk of the arsenic, mercury and lead found in the waste stream, while the bulk of the cadmium comes from nickel cadmium batteries (64.2%).

Table 1 Distribution of Trace Metals in MSW (Burnaby, BC, 1991)

Component	Amount in Waste Stream (%)	Metal Concentration as Percent of Total in Waste			
		Arsenic	Cadmium	Lead	Mercury
Paper	38.5	4.5	1.9	20.0	31.3
Plastic	8.2	0.6	14.7	15.7	2.2
Organics	34.5	40.2	7.8	41.5	35.4
Metals	2.3	27.7	5.7	3.2	11.5
Glass	1.8	0.5	0.6	1.1	0.5
Inorganics	0.7	0.5	0.9	5.7	0.3
Fines	7.5	6.4	2.5	12.0	14.6
Distribution in Incineration System		Arsenic	Cadmium	Lead	Mercury
Bottom Ash	78.6	50	3	70	0
APC Residue	13.3	45	92	28	0
Boiler/Economizer	8.0	4.9	4.8	1.9	40
Stack	0.1	0.1	0.2	0.1	60

Much of the contamination in the paper stream was found in the residual mixed paper fraction. Because this fraction was intimately mixed with the fines during the separation process, it could be highly contaminated with materials not inherent in the component itself and the results may illustrate cross-contamination of co-mingled materials. It is difficult to distinguish between the amount of these elements contributed by the residual paper itself and the amount associated with materials sticking to the paper. Cross-contamination also could occur for plastic materials; however, the absence of porous surfaces on plastics when compared to paper suggests that the possibility is less important for these components.

During the experiments in Burnaby, the median amount of cadmium entering the incinerator was 13.5 g/Mg with a range of 3.75 to 143 g/Mg on an as-sampled basis. Lead levels were relatively consistent being between 130 and 190 g/Mg of waste with a median of 163 g/Mg. The median concentration of mercury during the test program was 0.733 g/Mg with a range of 0.531 to 1.445 g/Mg on an as-sampled basis. The median arsenic feed rate was 7.8 g/Mg with the range being 3.9 to 13.9 g/Mg.

Partitioning in the System

From the foregoing it is evident that the trace metals found in waste must be found in the incineration process, either as part of the solid residues or ash produced or in the stack emissions. The question that remains is where would they most likely occur?

Exposure of the waste components to temperatures up to 1,000°C generally results in preferential partitioning of elements between the bottom ash and APC residues based upon the speciation of the particular element in the waste stream and the incinerator operating conditions. Lithophilic elements have boiling points in excess of 1,500°C and consequently remain in the bottom ash. Beryllium, magnesium, calcium, barium, chromium, manganese, iron, nickel, copper and aluminum would be expected to behave in this manner. On the other hand, the more volatile elements such as mercury, cadmium, tin, zinc, lead, arsenic and antimony would be expected to be transferred into the vapour phase during combustion. They typically partition between both the bottom ash and the APC residues depending upon speciation.

Table 1 illustrates the average distribution of the four trace metals discussed previously[2]. The mercury released through the stack represents the amount found in uncontrolled facilities with the APC filter capturing 40%. As discussed later, the addition of powdered activated carbon reduces mercury emissions to less than 5% of the input values.

Lead presents an interesting case. The partitioning of metals in the furnace is a function of the speciation of the metal and the vapour pressure of the compound. In the combustion zone of the furnace, lead and lead oxide are most likely to form $PbCl_2$ which is much more volatile than lead or lead oxide. It would thus be expected that lead would be vaporized and transported through the furnace to a point where the temperature was reduced and the material condensed back to a particulate form. Under this scenario, all the lead should be found at the end of the system, namely in the APC residue and the stack emissions. In fact, approximately 70% of the lead is normally found in the bottom ash and grate siftings and 28.3% was found in the fabric filter residue. Less than 0.1% of the lead was emitted to the atmosphere. Some lead clearly follows the thermodynamic pathway suggested by the Gibbs Free Energy considerations, other materials do not. This suggests that care must be taken in assigning causality to the concentration of any heavy metals found in various parts of the furnace.

With the caution of the previous paragraph in mind, generally the partitioning of the persistent toxics in the incineration system is predominantly towards the APC residue stream. This stream represents less than 15% of the total residues remaining after incineration, or less than 2% of the initial MSW mass charged to the furnace. While it contains concentrated materials, its small volume means that it can be effectively contained and its disposal carefully controlled.

The quality of the air pollution control system will dictate how much of the material partitioning to the gas stream and moving through the system will be captured. The question that remains it, how much will be released to the atmosphere and what can potentially migrate to the soils from the residues that are disposed. Those are the issues addressed in the balance of this paper.

APC Control Systems for MSW Incinerator

The material in the flue gas stream leaving the heat recovery system of a modern MSW incinerator consists of:

- gaseous products of combustion including carbon dioxide, hydrogen chloride, sulphur dioxide and oxides of nitrogen;
- vapour forms of metals and organics; and,
- solid particulate matter that includes organics and metals.

Testing programs on production installations have demonstrated that effective control of the emissions involves a combination of two processes:

- combustion control to minimize flue gas concentrations of conventional and trace organic contaminants; and,
- post-combustion control to reduce the amount of material leaving the stack.

While good combustion control is the corner stone of most MSW incineration regulations issued since 1987 and has become standard in all new facilities, it addresses concerns about trace organic pollutants. More important in the context of this presentation is post-combustion control or the performance of the air pollution control system.

Air pollution control [APC] system performance has improved dramatically over the last 15 years. Today, most new North American rely upon fabric filter technology for particulate control, but in Europe new installations may include various forms of wet scrubbers. Regardless of the system selected, APC systems remove unwanted contaminants such as trace metals and various acid gases from the flue gas stream exiting the facility. They rely upon both physical and chemical unit processes involving different solids removal and chemical conversion steps to effect control of unwanted emissions. The three key aspects of APC systems are reagent addition, temperature control, and particulate removal.

Conventional wet scrubber systems are effective at removing particles down to 1 μm. In Europe they are usually combined with particulate removal devices located upstream of the wet scrubbers so their fine particulate control performance is not detrimental to the overall system performance. The removal of smaller particles requires either the electrostatic precipitator [ESP]; or the fabric filter [FF].

The ESP is efficient in the collection of material in the 0.1 to 10 micron (μm) size range, although it does have some limitations in the 0.1-1.0 μm range. Removing finer material requires the use of extra fields. High efficiency ESP installations will remove 100% of the >50 μm material, >99% of the 5 μm material, and 98% of the 1 μm and smaller material. ESP's are generally considered to remove a fixed fraction of the material in the flue gas stream so if the inlet concentration is variable, it is likely the outlet concentration will vary too. ESP units have limited capabilities to remove pollutants in vapour form. In addition, because trace metallic and organic species tend to condense on surfaces, if the ESP has limited fine particle control the removal effectiveness for both metallic and organic species condensed on the surfaces of fine particles will be reduced.

Fabric filters are effective for particle sizes in the sub-micron range and their removal efficiency varies little with changes in inlet concentration. Test data from various operating facilities indicate that the current performance specifications for fabric filters on MSW incinerators are capable of maintaining emission concentrations below 15 mg/Rm^3 (dry gas at 25°C, 101.3 kPa, and 11% O_2).

It is not unusual for the bags to retain 10-20 times as much material as they filter out during any given filter cycle. This property is beneficial when APC systems rely upon chemical and physical transformations for pollutant control. Firstly, the presence of the cake on the bags enhances the removal efficiency of the system because it improves the interception of particles. Secondly, because fabric filters at MSW incinerators are frequently used in conjunction with lime addition for acid gas control, the material on the bags contained unreacted lime and provides additional polishing of the gas stream for acid gas removal.

The introduction of a finely divided reagent into the gas to control acid emissions provides an additional benefit in the removal of trace metallic and trace organic species. Volatile metals and organics demonstrate a high affinity for surface sorption and providing a large surface area in the form of fine particles in the APC system increases the probability of surface reactions. Thus modern APC systems have reduced emissions of volatile species, provided the particulate matter is well controlled. Temperature is an important aspect of acid gas control and because the adsorption effects for trace metals and organics are temperature dependent too, the systems provide improved performance at the normal operating temperature of 160°C.

Regardless of particulate matter control device or operating temperatures, mercury removal efficiencies are fairly low unless additional removal measures are employed. This metal has a low vapour pressure and mercury compounds can remain in the gas phase at temperatures less than 180°C. Since temperature control alone will not provide sufficient mercury control, reagents are added to the system to trap mercury. In Europe, wet systems utilize TMT to trap the mercury and remove it from the gas stream. In some combined systems, sodium sulphide has been injected into the APC but this has been replaced by activated carbon injection to reduce the risks to plant workers who were handling the sorbent (Guest and Knizek, 1991). Activated carbon appears to be a very effective mercury sorbent and improve removal efficiencies to the high 90% range. An additional benefit reported in some activated carbon injection test data is an increase in PCDD/PCDF removal rates.

Thus, the modern MSW incinerator's air pollution control system consists of a gas temperature reducing stage, the addition of lime or another acid gas reagent and powdered activated carbon to control mercury and PCDD/F emissions followed by a high performance particulate control device.

APC System Performance and Measurement Limitations

One of the most extensive Canadian data sets for MSW incinerator emissions has been developed at the KMS Peel Resource Recovery Facility located near Toronto International Airport in Brampton Ontario. Formerly known as the Peel Resource Recovery Inc., (PRRI), facility, the project converts 400 Mg/d of domestic, commercial and solid non-hazardous industrial waste into steam and then electrical energy. Provision is made in the plant via two parallel NRT separation lines to process the as-received waste to remove ferrous and non-ferrous metals, and glass/grit and other fine materials from the waste stream before it is fed to the incinerators. Each NRT line consists of a rotating fuel homogenizer equipped with mechanisms to liberate the waste from the bags, permanent magnets to remove ferrous, and a series of lifters to trap and remove fines from the waste as it rotates and tumbles down the drum. Following the drum, aluminum is separated from the processed waste stream before the remaining material is discharge to output conveyors and transferred to the storage floor. Four 2-stage incinerators and associated waste heat recovery boilers rated at 11,530 kilograms steam/hour provide steam to a turbine-generator set having a nominal capacity of 9,000 kilowatts. Each furnace has a design capacity of 100 Mg/day of 12.8 MJ/kg processed MSW. The operating permit requires that the secondary chamber temperature be maintained above 1000°C at all times when waste is fed to the furnace. While the furnaces are equipped with natural gas burners to meet the startup and operating requirements of the permit, these burners were not in used during any of the testing conducted at the facility.

The exhaust from each boiler passes to a common hot gas duct and the air pollution control (APC) system. The hot gas duct splits the flow between two parallel APC trains consisting of a wet spray humidifier, venturi dry lime injection tower and pulse jet fabric filter. Each side of the APC system has a rated capacity matching the flow from three operating incinerators. After the initial year of operation, the system has only operated in this mode (three incinerators to one APC) for testing purposes.

In 1991, while developing the permit conditions for this facility, regulators required that the facility be tested while the furnaces were using both raw and processed MSW to ensure that the worst case emissions were measured. This testing approach suggested that there was some uncertainty about the actual effects of pre-processing. The facility was tested a total of 12 times between April 1992 and October 1995 under a number of different operating conditions. Subsequently, annual triplicate testing under normal operating conditions has been conducted. The data from the first 12 triplicate series of tests provides a valuable insight into the performance of MSW APC systems and the affect that waste variations might have on emissions.

The facility's operating permit specifies emission limits for:

- total suspended particulate matter (20 mg/Rm^3 @ 11% O_2 3 test average);
- HCl (30 ppmdv or 50 mg/Rm^3 @ 11% O_2 - 24 hour average from CEM system);
- opacity (less than 5% - 2 hour average; and 10% for 6 minute average); and,
- PCDD/F (0.5 ng TEQ/Rm^3 @ 11% O_2 based upon a 3 test average).

In addition, there are operating windows for oxygen, carbon monoxide, total hydrocarbons and temperatures throughout the system. The plant continuously meets all these standards and also meets the point of impingement standards set out in Ontario Regulation 346. In effect, the latter imposes emission limits for many elements but testing normally includes only those elements included in typical laboratory metal scan lists. Emissions are deemed to be acceptable if the ground level concentrations, referred to as Point of Impingement (POI) levels, are met. For the PRRI facility the maximum POI occurs within 2 km of the stack. POI values are determined using a prescribed mathematical formula (simplified dispersion modelling) given in the Regulation.

In addition to the Certificate of Approval standards, the Ministry of the Environment in re-instituting the acceptability of EFW as a waste management option in December 1995, has recommended that new facilities meet stringent standards for lead, cadmium, and mercury and tighter limits on PCDD/F. The standards are: Cd - 14 μg/Rm^3, Pb - 140 μg/Rm^3, Hg – 57 μg/Rm^3 and PCDD/F - 0.14 ng/Rm^3.

The annual stack testing is required by the Ontario Ministry of Environment (MOE) to prove that the facility continues to meet the emission limits. This testing is funded by the Regional Municipality of Peel and the work reported in this paper was conducted under the direction of an independent consultant hired by the Region. Such testing has included measurements of the performance of each baghouse separately at both: maximum air to cloth ratios (flow from three units routed to one APC train) and, typical air to cloth ratios (both in the single APC train operation case of flow from two units routed to one APC train; and, in the normal mode of operation of all four incinerators on-line venting through both APC trains). Of the 10 series of tests reported herein, the first 6 were conducted at higher than normal air to cloth ratios in the fabric filter (three incinerators discharge to one of the two APC trains). The last four series were at normal air to cloth ratios (either both fabric filters were on line and all combustors were operating or one APC was off line and only two incinerators were operating). In addition, the MOE has required testing with both as-received MSW and the NRT processed waste to determine if there is a difference in emissions characteristics and also different seasons to ensure operation is the same in other periods.

Each triplicate series of tests were conducted at a specific test condition. Permit conditions with respect to operating temperatures were required to be met in all cases. While the flow rates in the stack were seen to change between the various test series these changes were the result of different APC operating conditions, not large swings in waste feed, heating value, or excess air levels. Furnace feed rate was in excess of 90% of the design heat input rate at all times. Typically, the run average oxygen content of the gas leaving the boiler was in the range of 7.5 to 8.5% as measured by the facility instrumentation. No significant changes in performance were registered during any of the test programs. Bottom ash quality during the initial testing program showed loss on ignition values below 5%. Subjectively, this burnout quality has been maintained throughout the life of the facility as the monthly ash shipments correlate well with the total amount of waste received at the facility.

A total of 27 individual runs are available for most parameters. These data can be used to identify changes in the air emissions induced by separation activities as well as seasonal differences in the waste stream.

Particulate matter, metals, PCDD/F and other trace organic emissions were measured during the stack testing campaigns: starting immediately following commissioning in May, 1992; seasonal testing in December, 1992 and March, 1993; and, subsequent annual testing in September 1993; October, 1994, November, 1995, October, 1996 and October, 1997. The latter two test series are currently not included in the data in this paper because they have not been compiled in the same format although subjectively the results were the same.

For each test series, a pre-test plan was submitted to the MOEE for approval. Test methods included:

- EPA Method 29 or its predecessor - California Air Resources Board (CARB) Method 436 - for particulate matter, fluoride and metals with the exception of mercury (during the first campaign of 9 tests, CARB Method 436 was also used for mercury determinations in parallel with the EPA Method 101A). The stack samples have been analyzed for up to 30 different metals over the course testing: (Ag, Al, As, B, Ba, Be, Ca, Cd, Co, Cr, Cu, Fe, Hg, Li, Mg, Mn, Mo, Na, Ni, P, Pb, Sb, Se, Sn, St, Te, Ti, V, Zn, and Zr). After the initial campaign, which included 9 test runs, failed to detect either Te or Zr, these elements were dropped from the analytical list for subsequent testing.
- EPA Method 101A for mercury; and,
- Modified Method 5 train as described in EPA Method 23 for trace organics. The Method 23 recovery and analysis methods were extended to allow for determination of PAHs, chlorobenzenes, chlorophenols and PCBs in addition to PCDD/F.

All isokinetic stack sampling was conducted for 240 minutes. In all cases, three-run averages were compared to the permitted emission concentrations. The highest value from each triplicate was used to determine the POI values since these are short term limits.

Table 2 provides summary data for the 15 metals on the US federal regulatory lists (US 40 CFR 261 (BIF Rule Regulations) and Clean Air Act, Section 112.) and the other metals analyzed that had significant responses during a statistical analyses that examined various in emissions due to operating conditions[5].

Negative values in the data table indicate the below detection limit values for 4 hours of sampling (about 7 dscm sample volume). Initial evaluation showed that using the detection

TABLE 2: Summary of KMS Peel Trace Metal and Acid Gas Data (μg/Rm³ @ 11% unless noted)

Sample Series	APC #	Waste Type	TSP (mg)	NOx (ppmdv)	SOx (ppmdv)	Hg 101a	Hg Meth29	Sb	As	Ba	Be	B	Cd	Cr	Co	Cu	Pb	Mg	Mn	Ni	P	Sb	Ag	Na	Sn
1	2	1	10.6	292	42.6	493	128	-0.3	-0.3	18.5	-0.1	184	504	1.0	-2.4	5.5	36.4	84.0	95.2	2.2	85.0	0.3	-0.5	389	9.0
1	2	1	1.5	264	32.7	346	242	-0.3	-0.3	1.5	-0.1	191	205	1.2	-2.4	5.8	4.1	2.3	107	2.1	133.1	0.2	-0.5	235	12.3
1	2	1	1.3	272	51.0	337	276	-0.3	-0.3	0.9	-0.1	207	51.6	1.0	-2.4	2.6	2.8	-1.1	1.5	5.7	17.5	0.3	0.9	175	12.3
2	1	1	3.0	268	45.1	490	337	-0.3	-0.3	1.0	-0.1	151	31.9	2.1	-2.4	3.5	2.5	-1.1	12.7	6.8	107.9	0.3	1.5	195	19.5
2	1	1	2.2	291	35.9	0	370	-0.3	-0.3	1.2	-0.1	188	8.9	2.0	-2.4	6.4	1.8	8.9	30.9	3.3	104.5	0.2	1.2	166	-1.1
2	1	1	2.9	315	40.6	363	382	-0.3	-0.3	1.4	-0.1	218	2.8	2.4	-2.4	4.2	2.1	10.6	4.8	3.9	130.8	-0.3	319	202	2.8
2	1	6				341																			
2	1	6				276																			
2	1	6				317																			
2	1	6				524																			
3	1	6	5.3	263	0.5	478	155	-0.3	-0.3	1.8	-0.1	88	0.8	75.9	-2.4	418	4.4	227	44.3	538	6.0	-0.3	1.3	297	2.5
3	1	6	6.8	287	15.3	477	219	-0.3	-0.3	1.5	-0.1	130	0.3	38.7	3.9	139	2.2	38.7	7.7	183	142.2	0.2	-0.5	180	5.5
3	1	6	5.3	260	28.7	456	438	-0.3	-0.3	1.0	-0.1	217	0.5	3.2	1.4	10.0	15.8	36.8	1.0	5.4	239.5	0.3	-0.5	245	99.6
5	1	1	15.0	238	15.8	113		0.1	0.1	0.4	-0.4	167	0.7	24.7	0.7	5.4	7.4	51.8	14.5	64.1	69.3	2.0	-4.2	764	8.9
5	1	1	2.7	245	18.6	221		0.1	0.0	-0.4	-0.4	171	1.8	5.2	-0.3	1.9	5.2	14.2	2.7	9.9	64.7	1.7	-4.2	455	8.2
5	1	1	8.1	223	15.5	163		0.1	-0.0	-0.4	-0.4	187	0.6	9.0	5.6	2.1	6.6	12.4	1.7	7.3	90.0	1.2	-4.2	559	8.0
6	2	1	3.8	200	3.8	183		0.0	0.2	0.6	-0.4	202	0.7	8.7	-0.3	1.8	5.2	11.7	75.8	0.5	44.0	0.5	-5.7	312	9.1
6	2	1	2.2	218	3.0	168		0.0	0.1	0.6	-0.4	175	0.5	13.6	0.3	1.7	6.0	8.1	15.7	10.1	73.0	0.5	-5.7	199	6.0
6	2	1	1.9	228	7.9	331		0.0	0.1	0.6	-0.4	213	0.5	2.6	-0.3	0.8	4.9	5.6	4.4	1.4	73.9	0.5	-5.7	191	9.1
7	3	1	1.6	241	5.0	308		0.1	0.2	1.6	-0.2	226	0.6	1.0	-1.0	5.5	4.7	6.4	2.1	-1.9	10.2	-0.5	2.3	389	6.1
7	3	1	1.1	208	5.1	321		0.1	0.2	0.9	-0.2	181	0.6	-0.6	-1.1	5.4	-4.8	28.7	2.0	2.2	46.1	0.6	2.4	367	15.7
7	3	1	5.4	298	11.6	409		0.1	0.2	0.7	-0.2	174	0.7	2.2	-4.9	6.9	4.6	29.1	3.3	9.1	42.3	0.2	2.3	389	13.2
8	3	6	2.7	96	2.3	218		0.1	0.2	0.7	-0.2	159	0.9	1.8	-0.9	5.9	5.6	27.3	1.8	2.4	36.5	0.3	2.0	367	11.7
8	3	6	2.7	226	11.9	266		0.1	0.2	0.6	-0.2	159	0.9	1.1	-1.0	5.1	7.8	24.5	0.8	-1.9	31.5	-0.2	2.2	280	11.0
8	3	6	2.1	221	3.7	250		0.1	0.2	0.6	-0.2	190	0.7	2.9	1.2	5.8	4.4	17.9	2.7	4.6	20.1	-0.2	2.2	407	11.1
9	1	1	10.0	231	9.9	480		2.4	-1.5	4.6	-0.5	74	4.0	4.8	-2.3	6.2	43.8	50.0	31.2	30.0	21.7	5.5	-4.8	1500	-11.5
9	1	1	3.5	230	8.5	420		2.0	-0.9	2.2	-0.3	145	3.7	3.6	1.1	5.7	32.1	32.8	5.2	25.2	-11.7	1.1	-2.7	832	-6.5
9	1	1	4.1	235	16.5	240		2.0	-0.8	2.0	-0.3	226	2.9	5.7	1.3	5.1	26.7	38.6	7.7	305	-11.1	1.2	-2.6	862	-6.5
10	2	1	3.1	241	12.1	270		-0.9	-0.8	1.7	-0.3	211	1.9	9.2	21.0	3.2	20.5	20.1	26.5	7.7	-10.8	1.0	2.5	684	-6.1
10	2	1	2.7	234	16.3	310		-0.8	-0.8	1.0	-0.3	99	1.2	8.9	1.2	2.2	14.2	13.8	3.1	19.4	11.5	0.9	-2.6	398	-6.2
10	2	1	0.5	227	16.7	280		-0.8	-0.8	1.5	-0.3	155	1.6	3.7	1.0	2.5	18.7	16.9	1.7	6.3	-10.6	1.0	-2.5	738	-6.0

NOTE: Negative signs denote values that were less than the analytical detection limit.
Series numbers designate test series (no metals testing during series 4)
Waste Type Key 1 = MSW 6 = NRT APC Designator: 1 = north 2 = south 3 = both

limit [DL] to characterize BDL results produced many false positive conclusions. This occurred because the DL is not an indicator of the actual concentration, rather it mirrored trends in analytical method and these moved by chance with the operating conditions, thereby confusing the interpretation. To avoid this problem, below detection limit values were eliminated from further consideration.

Table 3 provides the blank metals train data for the various test series. These values result from assembling and recovering an unused sampling train and converting the blank train analytical values into apparent concentrations using the average sample volume and oxygen level associated with a triplicate test series. In many cases the stack data is of the same order of magnitude as the blank train value suggesting that either the source is very clean or the method seriously flawed.

TABLE 3: Summary of KMS Peel Blank Train Data for Sampling Runs (μg/Rm³ @ 11% unless noted)

Sample Series	APC #	Waste Type	TSP (mg)	NOx (ppmdv)	SOx (ppmdv)	Hg 101a	Sb	As	Ba	Be	B	Cd	Cr	Co	Cu	Pb	Mg	Mn	Ni	P	Sb	Ag	Na	Sn
2	1	6	10.6				-0.2	-0.2	0.9	-0.3	9.0	-0.5	-0.9	-2.2	2.8	-5.1	14.6	-1.1	-2.5	-14.4	-0.2	-2.6	155.	-21.8
3	1	6	1.5				-0.2	-0.2	0.9	-0.3	9.6	-0.5	-1.0	-2.4	4.9	-5.4	15.5	1.9	-2.6	29.1	-0.2	-2.7	101.	-23.2
5	1	1	1.3				0.1	0.1	0.2	0.0	10.3	0.8	0.7	1.7	0.3	1.8	2.2	0.4	1.5	0.0	0.0	-2.2	29.7	2.1
6	2	1	3.0				0.0	0.1	0.6	0.0	3.9	0.4	0.4	0.4	0.3	0.6	1.2	1.7	0.9	3.2	0.1	-3.0	11.8	0.4
7	3	1	2.2				0.1	0.1	0.6	-0.2	4.1	-0.4	0.5	-1.0	4.9	-4.3	10.7	0.8	-1.8	11.2	0.1	-2.1	159.	7.4
8	3	6	2.9				0.1	0.1	0.8	-0.2	5.7	-0.5	0.7	-1.1	4.8	-4.6	20.5	3.0	-2.0	23.7	0.1	-2.3	105.	9.8
9	1	1					-0.9	-0.9	1.7	-0.3	13.8	-0.6	1.2	-1.1	1.0	11.5	12.0	2.7	32.1	-12.3	1.0	-2.9	676.	-6.9
10	2	1	0.5				-0.8	-0.8	1.2	-0.2	6.0	-0.5	0.5	-0.9	0.6	10.1	6.0	0.7	3.1	-10.1	0.8	-2.4	336.	-5.6

NOTE: Negative signs denote values that were less than the analytical detection limit.
Series numbers designate test series (no metals testing during series 4)
Waste Type Key 1 = MSW 6 = NRT APC Designator: 1 = north 2 = south 3 = both

The results of the KMS Peel testing and initiatives in various jurisdictions to lower emission levels even further than suggested by the 1995 Ontario guidelines prompted a comparison of test results in December, 1996. During emission control system testing at an ESP equipped MSW incinerator, dual- and quad-train stack sampling was conducted to establish the level at which US EPA Method 29 imprecision is no worse than ± 30%[6]. This level is between 34 and 82 mg/Rm³ for particulate; 29 and 51 μg/Rm³ for Cd; 370 and 950 μg/Rm³ for

Pb; and, 36 and 51 μg/Rm^3 for Hg all at 11% O_2. These values are in the middle of the measured emission values for this particular source. Given that an ESP is less capable of meeting low emission levels than a fabric filter equipped APC system, the fact that the dual train values are higher than seen in the KMS Peel data should not be surprising. The implication of the dual train test results is that at the low end results could have significantly more than 30% imprecision. Therefore actual emissions could be lower, or higher than the reported numbers. With this imprecision, it is not surprising that KMS Peel blank train results are in the same range as performance test results. More importantly, the results suggest accurately characterising emissions at extremely low levels may not be possible with the current manual sampling methods. Any sources that consistently test at levels at or below those shown above should be considered to be operating properly. The most important information that can be gained from repeated testing of such sources is the potential to determine when performances has deteriorated to a level that requires intervention. For instance, given that the bags in fabric filter system wear with time, increases in the particulate emission levels, or indeed in a particular metallic species should prompt an examination of maintenance procedures. Either the bags are past their useful life span, or maintenance activities have not had the desired effect. If bag life is not in question, re-testing may find that the results were an anomaly.

Given that the KMS Peel results are routinely duplicated every time the facility is tested, it can be reasonably assumed that they represent typical performance. Absolute concentration data is difficult for many people to interpret, and indeed, many sources are ranked upon their annual emissions. If the mean tested emission value were assumed to equal the yearly average, a reasonable assumption given the probability of getting both high and low values at any time during the year including during testing, the annual emission reduction can be calculated. Geometric mean values of the data were calculated to determine emission averages because the data have been found to be log normally distributed. The calculation of annual emissions in Table 4 uses the average flow for full operation, 4 furnaces on line, of 90,216 Rm^3/hour; the number of hours per year this would occur based upon an annual capacity factor of 90% of the full design load of 400 Mg/day with operation year round. The annual emissions in Mg/year for both the existing case and the case where the facility complied with the new Ontario guidelines are provided in the table.

Several observations are appropriate. Firstly, the facility, as it is currently operating, meets the new target emission levels for lead, cadmium and total suspended particulate matter. Both the mercury and PCDD/F emissions have dropped over the years since the facility commenced operation. While the addition of PAC will lower these emissions, the stated reduction in annual emissions may be an over-estimate. On the other hand, the projection of improved removal efficiency may be conservative and the reduction could be more significant. Overall, the facility will release about 2300 kg/year of particulate matter; after the addition of PAC less than 41 kg/year of mercury will be emitted. Lead and cadmium emissions will be less than 5 and 2 kg/year respectively. Comparing the latter two values to those reported for 1995 on the NPRI lists, they are very small.

The Availability of Trace Metals in the Residue Streams

The WASTE Program[2] undertook leachate extraction tests on selected waste and residue components to identify the potential mobility of trace elements found in the MSW. The testing used the Sequential Chemical Extraction Procedure to assess the potential solubility of trace metals under various landfill regimes. This sequential batch leaching protocol employs increasingly aggressive leachant in each of the five steps. Step A solubilizes the ion exchangeable species likely to be released on initial contact with water; B represents that

TABLE 4: Summary of KMS Peel Resource Recovery Inc. Emission Data (μg/Rm³ @ 11% unless noted)

Test Date	Sample Series	APC #	Waste Type	On-line	Moisture %	Oxyge %	Temp F	Flow Rm³/hr	Pb	Cd	Hg	TSP mg	NOx ppmdv	SOx ppmdv	Sum PCDD/F ng	I-TEQ Cdn ng	Cr	Cu	Mg	Ni	Zn
920505	1	2	1	3	14.4	12.5	332	72037	36.39	503.91	493	10.6	291.6	42.6	9.76	0.91	1.01	5.49	83.99	2.24	15.26
920506	1	2	1	3	15.7	12.5	327	58615	4.11	205.41	346	1.5	263.8	32.7	4.72	0.36	1.17	5.75	2.35	2.05	11.06
920507	1	2	1	3	17.0	13.5	335	69489	2.84	51.64	337	1.3	272.4	51.0	4.38	0.30	1.04	2.61	1.10	5.69	7.11
920508	2	1	1	3	15.3	14	330	65411	2.53	31.93	490	3	267.9	45.1	7.36	0.40	2.08	3.55	1.10	6.84	8.87
920509	2	1	1	3	16.9	12	325	67790	1.78	8.88	363	2.2	291.0	35.9	6.71	0.38	2.01	6.36	8.88	3.25	29.29
920510	2	1	1	3	16.5	11	334	65071	2.12	2.83	341	2.9	315.3	40.6	2.59	0.18	2.41	4.25	10.62	3.90	19.83
920520	3	1	6	3	17.7	11	331	64562	4.43	0.82	276	5.3	263.5	0.5	3.31	0.32	75.94	418.31	226.88	537.92	50.50
920521	3	1	6	3	16.6	12.5	336	64902	2.21	0.33	317	6.8	286.7	15.3	3.48	0.38	38.74	139.45	38.74	182.61	19.92
920522	3	1	6	3	17.5	12.5	321	63203	15.77	0.50	524	5.3	260.5	28.7	2.69	0.36	3.15	9.96	36.84	5.39	4.98
920809	4	2	1	3	17.7	11.0	325	60299							1.02	0.11					
920810	4	2	1	3	20.0	10.6	326	57407							1.39	0.16					
920811	4	2	1	3	21.0	10.1	328	56075							1.03	0.10					
921214	5	1	1	3	18.0	10.9	322	71867	7.39	0.74	478	15	237.6	15.8	18.67	0.70	24.65	5.42	51.76	64.09	27.36
921215	5	1	1	3	16.9	10.8	314	71698	5.22	1.79	477	2.7	245.0	18.6	11.16	0.32	5.22	1.86	14.17	9.95	18.15
921216	5	1	1	3	18.2	10.7	312	71698	6.57	0.58	456	8.1	223.2	15.5	3.94	0.38	9.00	2.12	12.40	7.29	23.83
930319	6	2	1	3	17.8	10.5	319	76455	5.20	0.71	113	3.8	199.9	3.8	10.58	0.67	8.66	1.82	11.70	0.45	49.82
930320	6	2	1	3	18.0	10.9	323	77814	6.03	0.54	221	2.2	217.8	3.0	6.07	0.37	13.63	1.68	8.05	10.06	17.43
930321	6	2	1	3	17.8	10.9	323	75265	4.88	0.49	163	1.9	227.7	7.9	10.17	0.35	2.56	0.77	5.58	1.39	13.24
930913	7	3	1	4	19.4	10.2	322	92595	4.65	0.57	183	1.6	240.6	5.0	3.67	0.36	0.97	5.53	6.40	1.90	15.39
930915	7	3	1	4	22.6	9.4	318	76964	4.80	0.59	168	1.1	208.3	5.1	2.34	0.49	0.60	5.37	28.70	2.18	8.15
930920	7	3	1	4	20.8	10	322	89367	4.58	0.67	331	5.4	297.9	11.6	1.51	0.16	2.19	6.87	29.08	9.06	0.58
930916	8	3	6	4	19.0	9.2	321	92935	5.62	0.91	308	2.7	95.6	2.3	1.76	0.13	1.75	5.93	27.32	2.39	19.20
930917	8	3	6	4	20.4	10	323	93954	7.76	0.94	321	2.7	226.2	11.9	1.31	0.12	1.05	5.06	24.46	1.90	17.37
930918	8	3	6	4	19.3	9.9	321	95484	4.43	0.66	409	2.1	220.5	3.7	1.64	0.17	2.93	5.79	17.89	4.60	14.51
941018	9	1	1	2	16.2	11.9	323	54708	43.75	4.04	218	10	231.0	9.9	4.93	0.46	4.75	6.21	49.99	30.00	146.65
941019	9	1	1	2	17.4	11.6	323	51310	32.11	3.69	266	3.5	229.9	8.5	2.18	0.17	3.59	5.68	32.82	25.21	575.30
941020	9	1	1	2	15.8	11.9	322	54368	26.67	2.92	250	4.1	235.4	16.5	1.36	0.15	5.67	5.10	38.57	304.76	115.19
941024	10	2	1	2	13.6	11.9	318	56237	20.51	1.92	340	3.1	240.9	12.1	1.36	0.15	9.21	3.24	20.07	7.72	80.67
941025	10	2	1	2	15.9	11.8	318	54028	14.24	1.22	320	2.7	233.9	16.3	1.28	0.15	8.90	2.24	13.78	19.43	46.93
941026	10	2	1	2	16.1	11.4	320	53926	18.66	1.56	180	0.52	227.2	16.7	1.72	0.23	3.68	2.50	16.92	6.29	41.66
ANNUAL EMISSIONS																					
EXISTING SITUATION																					
Average Full Flow Load					20.3	9.8	321.2	90216.6													
Mean Concentration					17.5	11.2	323.8		7.1	2.2	300.4	3.2	238.1	11.6	3.2	0.27	3.9	5.2	16.4	8.3	22.2
Emissions, Mg/annum									0.005	0.002	0.214	2.3	318.5	21.6	2.27E-06	1.91E-07	0.003	0.004	0.012	0.006	0.016
IMPLICATION OF COMPLIANCE WITH A-7*																					
Average Full Flow Load									142.00	14.00	57.0	17.0	110.0	21.0	NA	0.14					
Mean Concentration					17.5	11.2	323.8	0.0	7.1	2.2	57.0	3.2	110.0	11.6	1.7	0.14	3.9	5.2	16.4	8.3	22.2
Emissions, Mg/annum									0.005	0.002	0.041	2.3	147.1	21.6	1.19E-06	9.996E-08	0.003	0.004	0.012	0.006	0.016
Reduction in Mg/Annum based upon 400 Mg/day feed									0	0	0.17	0	171.3	0	1.09E-06	9.121E-08	0	0	0	0	0

NOTES: Shaded values denote values that were less than the detection limit.
Series numbers designate test series (no metals testing during series 4)
* Changes in annual emissions that would be result if existing facility was brought into compliance with new standard; the only emission values that change are those associated with NOx, Hg and PCDD/F's and these are assumed to be at the new regulatory limit.

APC # Key 1 =north 2 = south 3 = both
Waste Type Key 1 = MSW 2 = NRT

material that would be solubilized during early municipal landfill; C represents metal ions likely to be released under the severe reducing conditions in older landfills; whereas the D and E represent the residual metal ions that are normally unavailable in nature. Looking upon A+B as short term available, and A+B+C as long term available, it is possible to rank the materials on the basis of the total amount of a trace element released in both time frames and provides some important insights into the potential for environmental release of the trace metals from waste management facilities.

The WASTE Program[2] provided some data to compare the availability of trace metals in incinerator residue streams with that in the components. Assuming that the APC residues would need to be treated, or handled in a manner to limit releases to the environment, the cumulative effect of releases from the residues was compared to that of the waste components using the median SCEP data by MSW component or residue group, and the median concentration data. Taking these factors into consideration, in the short term slightly more lead might be released from incineration[7], but in the long term, the amount released appears to decrease suggesting incineration may stabilize some of the lead in the waste stream. Data from spiking studies conducted in Burnaby[8] showed that cadmium leachability from all residue streams was reduced during the liquid spiking runs suggesting that the material was converted to an oxide form in the furnace. The cadmium leachability during pellet spiking showed the same behaviour as the liquid spiking tests; it was reduced.

The characteristics of the APC residues generated by the use of activated carbon are also of concern. The removal of additional mercury from the gas stream will raise the concentration of mercury in the APC residues. Limited testing has shown that the mercury removed on the activated carbon is fixed to this medium and less environmentally available.

With the environmental availability of materials being reduced through the combustion process, it has been suggested that the most effective way to control metals in MSW could be incineration followed by controlled disposal of incinerator residues[6].

CONCLUSIONS

Overall, the modern EFW facility does a good job of controlling trace metal emissions from the stack. High performance APC systems, supplemented with special reagents to control particularly problematic species, are installed in direct response to the extremely stringent performance standards placed upon these facilities.

While trace metals are present in our MSW streams, the concentration of these materials is unlikely to be eliminated by product formulation modifications because a considerable portion of the trace metal input to MSW comes from the natural environment where these elements originate.

The residues from the MSW incineration process appear to be less prone to releasing the trace metals than are the original components of the waste stream and thus the thermal treatment process would appear to offer some potential for stabilizing these materials.

Overall, thermal treatment of MSW would appear to reduce the availability of trace metal species in the environment.

REFERENCES

1. Chandler, A. J., and H.G. Rigo, 1993. Mercury Partitioning in Municipal Solid Waste - The Myth and Reality. A presentation at the 6th International Conference on Ash Management & Utilization (Ash VI), Arlington, Virginia. November

2. The WASTE Program, 1993.

3. Markham Study, 1995. A report on a Waste Diversion Pilot Study in the City of Markham, Ontario.

4. Chandler, A.J., 1996. State-of-the-Art Assessment of Municipal Waste Incineration. A report to the International Air Quality Advisory Board of the International Joint Commission. February.

5. Chandler, A.J., and H.G. Rigo, 1996. Waste Separation, Does it influence Municipal Waste Combustor Emissions? A paper presented at 1996 ASME Biennial Waste Processing Conference. Atlantic City NJ, April.

6. Rigo, H.G. and A.J. Chandler, 1997. Practical Quantitation Limits for Reference Methods 23, 26 and 29. Submitted for publication in the Journal of the Air and Waste Management Association.

7. Chandler, A.J., 1994. Issues with Respect to Lead in MSW. A Presentation for The OECD Risk Reduction Program for Lead, September, 1994 Toronto Workshop on Lead in Products.

8. Chandler, A.J., 1995. Cadmium in MSW Management Systems. A Presentation for the OECD Workshop on Cadmium, Stockholm, Sweden October.

Towards an improved natural sources inventory for mercury

P.E. Rasmussen
Geological Survey of Canada, Natural Resources Canada
601 Booth St., Ottawa, Ontario, Canada, K1A 0E8

G.C. Edwards, R.J. Kemp, and C.R. Fitzgerald-Hubble
School of Engineering, University of Guelph
Guelph, Ontario, Canada, N1G 2W1

W.H. Schroeder
Atmospheric Environment Service, Environment Canada
4905 Dufferin St., Downsview, Ontario, Canada, M3H 5T4

ABSTRACT

Estimates of the total emission of elemental mercury vapour (Hg^0) to the atmosphere from natural sources vary by orders of magnitude. The uncertainty arises from the scarcity of temporally and spatially representative data. This paper describes field studies conducted from July 1996 to September 1997 directed at developing and applying methodologies for measuring atmospheric Hg^0 fluxes from natural sources. Preliminary results for five contrasting geochemical settings in Canada and the USA indicate a positive relationship between Hg^0 flux and total Hg concentrations in the substrate

Proceedings of
Metals and the Environment: An International Symposium
Edited by Jim Skeaff
The Metallurgical Society of CIM
100th CIM Annual General Meeting, 1998

INTRODUCTION

Elemental mercury vapour (Hg^0) is released into the atmosphere from a wide variety of natural and anthropogenic emission sources. Existing estimates of natural rates of Hg^0 emissions are poorly constrained due to the scarcity of temporally and spatially representative flux data (Schroeder *et al.*, 1989; Rasmussen, 1994). As a result, estimated ratios of natural to anthropogenic emissions are very uncertain. Published estimates for air-surface exchange fluxes (and input-output budgets derived from environmental measurements or mass balance models) are still associated with undesirably large uncertainties (Schroeder *et al.*, 1989; Schroeder, 1996). It is thus essential to determine natural volatilization flux values directly under a variety of environmental conditions, to properly understand, in qualitative as well as quantitative terms, the biogeochemical cycling of mercury in the environment.

This paper outlines field studies conducted from July 1996 to September 1997 to develop and apply *in situ* flux measurement technologies in natural settings. Other journal papers are in preparation which discuss the method adaptations required for monitoring fluxes in complex terrains, the relative influence of meteorological factors, soil conditions (moisture and temperature) and geochemistry, and the cycling pathways through which mercury species are liberated from bedrock into the surface environment. The purpose of this paper is to discuss the relevance of our preliminary results in terms of improving one component of emissions inventories – quantification of the natural Hg^0 flux to the atmosphere from soils and rocks.

PREVIOUS INVENTORIES OF NATURAL MERCURY EMISSIONS

Information on natural emissions of mercury in North America was compiled in the late 1980s, as part of a larger study of emissions, production and usage of 14 priority toxic chemicals in the Great Lakes ecosystem sponsored by the International Joint Commission (Voldner and Ellenton, 1987; Voldner and Smith, 1989). Citing the *National Inventory of Natural Sources and Emissions of Mercury Compounds for Canada* (EPS, 1981), these documents reported an estimate for natural emissions of elemental mercury (Hg^0) in Canada of 3,500 tonnes per year (t/yr), with the chief sources being emissions from vegetation, outgassing of soils and rocks, and emissions from freshwater surfaces and wildfires. In the *National Inventory* itself, the authors stated that 3500 t/yr is only an "order-of-magnitude approximation," serving to provide a framework within which new data may be incorporated.

In the present study, sites with contrasting geochemical characteristics were selected in order to address certain hypotheses underlying the *National Inventory* estimate. Specifically, the *National Inventory* introduced multiplicative correction factors to account for two geological situations resulting in elevated concentrations of mercury in soil parent material: the presence of sulphide mineralization and the distribution of early Proterozoic (Aphebian) shales. Starting with an estimated Hg^0 flux between 0.008 and 0.014 $\mu g/m^2/hr$, depending on soil type and temperature, a four-fold correction factor was applied to 10 census regions in central British Columbia and a two-fold correction factor was applied to other mineralized areas of British Columbia, Alberta, Yukon, and parts of other provinces underlain by early Proterozoic shales (EPS, 1981). Thus, the *National Inventory* recognized that natural background concentrations in environmental materials are determined not just by the mercury content of local/nearby ore deposits, as is sometimes assumed, but rather by the presence of mercury in varying amounts in a number of common geological situations.

There were, however, very few measurements of mercury emissions from sulphides reported in the literature at the time the *National Inventory* was compiled, and there were no reported measurements of mercury emissions from early Proterozoic shales (or from shales of any age). Correction factors for mercury emissions from mineralized areas were extrapolated from the work of Lindberg *et al.* (1979) and references cited therein, and correction factors for mercury emissions from early Proterozoic shales were extrapolated from mercury concentration data of Cameron and Jonasson (1972).

Thus, the present study was designed to address a key underlying hypothesis of the *National Inventory* (EPS, 1981) - that Hg^0 emissions from soils with sulphides or early Proterozoic shales as the predominant soil parent material are significantly higher than emissions from other soils in Canada.

METHODOLOGY

Rationale for site selection

For the 1996 pilot study, sites were chosen in and around a natural mercury anomaly near Clyde Forks, Ontario (National Topographic Series (NTS) Map 31F/2), where the principal mercury-bearing minerals are tetrahedrite and cinnabar (Nikols, 1972). The Clyde Forks anomaly (located in Figure 1) was an appropriate choice for the initial method development work, as earlier work demonstrated that significant concentrations of Hg^0 vapour are present in soil gas drawn from 1 m below surface, due to the weathering of sulphide minerals at this site (Bristow, 1972). The reference to the mineralization at Clyde Forks as an "anomaly" indicates that Hg^0 emissions measured here are by no means representative of Precambrian Shield rocks and soils. The only major cinnabar (HgS) occurrences in Canada are located in British Columbia, although cinnabar is known to be associated with other sulphide deposits east of the Rockies, such as the Hemlo deposit in Ontario (Harris, 1989). At Clyde Forks, the cinnabar is associated with barium-copper-silver-antimony sulphide mineralization in a fault zone located in the Grenville Structural Province of the Precambrian Shield (Pauk, 1989). High mercury concentrations related to the sulphide mineralization are also found in the sediments of nearby lakes (Jonasson, 1976).

In 1997, two contrasting sites were selected in the Precambrian Shield west of Thunder Bay, Ontario (NTS map 52A/5): an early Proterozoic black shale site and an Archean metamorphosed granite site (Figure 1). A regional scale natural mercury anomaly in this area, attributed to two metal-enriched sedimentary rock formations (the Rove and Gunflint Formations), was identified previously through the analysis of lake sediments collected under the National Geochemical Reconnaissance (NGR) program (Friske and Coker, 1995, and references cited therein). The black shale site selected for this study is located within the Proterozoic Rove Formation, a carbonaceous shale unit which extends across the Canada-US border from Ontario into Minnesota (Geul, 1970). In Ontario, the anomaly covers an area of about 295 km^2 as determined by Hg concentrations in sediment samples collected from 22 lakes, out of which 21 lakes exceed Environment Canada's interim threshold sediment quality guideline of 174 ppb for freshwater sediment (Environment Canada, 1997; Rasmussen *et al.*, 1998). An exposed outcrop of granitic composition and low mercury concentration in Kakabeka Falls Provincial Park, located about 20 km away from the black shale site, was selected to provide the contrasting background site.

In September 1997, measurements were made at a site located about 15 km south of Reno, Nevada as part of an international Hg^0 flux intercomparison study (Figure 1). The study was sponsored by the Electric Power Research Institute (EPRI), and was organized and hosted by Dr. Mae Gustin, of the University of Nevada-Reno (UNR). The study site is located in the Steamboat Springs, Nevada, geothermal region and is naturally enriched in Hg. Due to unexpected rain events over part of the monitoring period, the study was dubbed "Desert

STORMS" (Desert Study and Tests Of the Release of Mercury from Soils). A description of the study may be found on the internet (http://tekran.com/nevada/nevada.html).

Orientation surveys

Preliminary soil geochemistry surveys were conducted at each study site to characterize background and anomalous total Hg concentrations in the substrate and to determine optimum locations for monitoring atmospheric Hg^0 fluxes. The "Clyde Forks anomaly" site (Figure 2) is characterized by undisturbed sparsely vegetated soil developed over the mineralized fault zone, and an average total Hg concentration in the B horizon of 125 ppm (s.d. 129; n = 5). The "Clyde Forks background" site (Figure 2) is similar in appearance and is located only 20 m away from the mineralized fault zone, but has an average total Hg concentration in the B horizon of 167 ppb (s.d. 55; n = 8), nearly three orders of magnitude lower than the undisturbed anomaly site. An orientation survey in the Thunder Bay area confirmed previous findings of Friske and Coker (1995; and references cited therein) that the Rove shale has anomalously high total Hg concentrations, averaging 845 ppb (s.d. 863; n = 6) compared to concentrations in the metamorphosed granite which are below the detection limit (10 ppb). For the Nevada geothermal site, preliminary data provided by M. Gustin (University of Nevada) indicate that total Hg concentrations in the substrate vary between 1 and 12 ppm. Detailed follow-up surveys were conducted to characterize substrate Hg concentrations at each of the "Thunder Bay granite," "Thunder Bay black shale," and "Nevada geothermal" sites (Figure 2). At the time of writing, the follow-up samples have been submitted for analysis, and only the preliminary survey data are available for the 1997 studies.

Mercury concentrations reported here as "total mercury" are determined by aqua regia digestion and cold vapour atomic absorption spectrometry (AAS), with a method detection limit of 10 ppb. The large relative standard deviations (RSD) reported here (often over 100% RSD) are due to the natural heterogeneity of Hg distribution in the soil and bedrock samples, rather than to analytical error. Analytical error is less than 10% as determined by the insertion of a blind CCRMP "Till-1" standard after every 10 samples. Due to the large within-site and within-sample variability at least three samples are required to obtain representative Hg concentration values at each of the flux chamber sites and more samples are required at the micrometeorological sites (determined by the size of the footprint).

Measurement of air-surface exchange of mercury

Both flux chamber and micrometeorological methods were used for the *in situ* measurement of the air-surface exchange of mercury in each study area. It is important to note that the flux can assume either a positive value which corresponds to an emission of Hg^0, or a negative value which corresponds to a deposition of Hg^0 (Schroeder, 1996). The chamber method offers the advantage of high sensitivity while the micrometeorological approach provides the advantage of non-invasive direct measurement of the mercury vapour flux. In general, the chamber method appears to yield lower flux values than the micrometeorological method under similar operating conditions, but at this time a direct comparison is difficult due to the different footprint areas (areas of substrate covered by each method) and the spatial heterogeneity of mercury concentrations within each site (often over 100% RSD).

The micrometeorological method is based on the work of Edwards *et al.* (1995), and the chamber approach is based on Schroeder *et al.* (1989). Both methods use a Tekran Model 2537A mercury vapour analyzer, which accurately determines total gaseous mercury in air even at concentrations below 1 ng/m^3. The instrument samples air and traps mercury vapour into a cartridge containing an ultra-pure gold adsorbent. The trapped mercury is then thermally

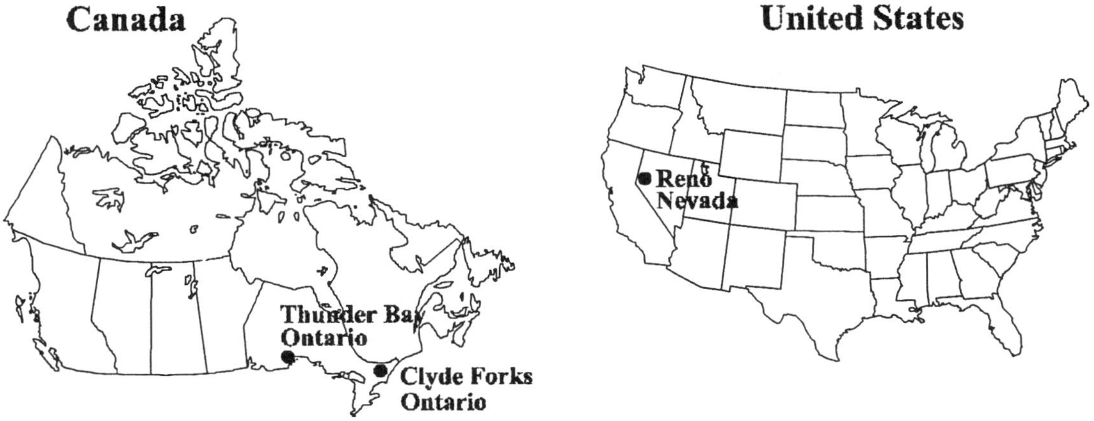

Figure 1. Location of field monitoring sites. Two sites ("granite" and "black shale") are located west of Thunder Bay, in northwestern Ontario and two sites ("sulphide anomaly" and "background") are located near the town of Clyde Forks, in southeastern Ontario. The "geothermal site" is the location of the Desert STORMS intercomparison study, located just south of Reno, Nevada.

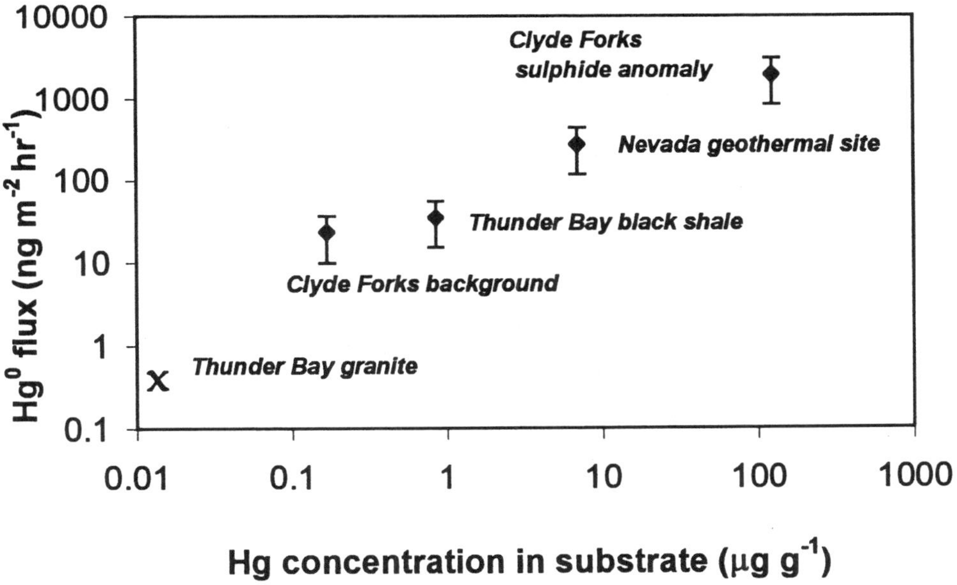

Figure 2. *In situ* field measurements of the rate of emission of Hg^0 to the atmosphere in contrasting geological settings under comparable temperature and rainfall conditions. Mercury concentration data (x-axis) are total Hg concentrations (μg/g or ppm) determined in soil or bedrock samples by aqua regia digestion and cold vapour AAS detection, and represent preliminary survey data only. Vertical error bars show standard error of the mean Hg^0 flux (y-axis) averaged over entire measurement period (night and day) using micrometeorological-gradient method (3.5 days in July 1996 at each of the Clyde Forks sites, for 18 days from July to August 1997 at Thunder Bay, and for 5 days in September 1997 at the Nevada site). The exception is the "granite" site (X) which is estimated using the chamber method as fluxes were below the detection limit of the micrometeorological method.

desorbed and detected using cold vapour atomic fluorescence spectrometry (CVAFS). A dual cartridge design allows alternate sampling and desorption, resulting in continuous measurement of the air stream.

Micrometeorological Method: The micrometeorological method is a gradient flux approach where the flux (F) is determined by the relationship

$$F = K * \Delta C / \Delta Z,$$

where K is the transfer coefficient, ΔC is the difference in mercury vapour concentration in air (ng/m^3) between 2 sampling heights (typically 10 cm and 30 cm) above the soil surface, and ΔZ is the distance between the two sampling heights (typically 20 cm). Thus the term "$\Delta C/\Delta Z$" refers to the gradient, which is determined by switching air concentration measurements between the upper ("up") and lower ("down") sampling levels. One Tekran mercury analyzer is used to determine ΔC. The up and down concentration measurements are subtracted sequentially and are smoothed to produce a running mean concentration difference. For ideal sites the transfer coefficient "K" is generally derived from a 5 cm sonic anemometer and a 4 cup wind profile. Similarity is assumed and corrections for atmospheric stability are applied to the momentum transfer coefficient. For complex sites the transfer coefficient may be based on additional measurements which include: additional turbulence data, energy balance data (sensible heat, latent heat, net radiation and storage), and additional measurements of scalars using gradient and eddy correlation.

A versatile gradient approach has been developed to allow for the method to be tailored to suit individual site characteristics and experimental requirements. The gradient intake heights are selected to look at a specific spatial average area (i.e., "footprint") of emission/deposition. This approach is suitable for looking at small or large footprints. The size of the footprint not only depends on the height of the intakes but on site geometry and atmospheric conditions. Typically the gradient measurement is set such that 90% of the flux measured originates from an upwind fetch of 5 to 15 meters.

The apparatus is portable and robust, and has been successfully deployed in a variety of rugged field environments operated on portable generator power. The gradient and micrometeorological apparatus are mounted on two short pipes designed to minimize aerodynamic interference. Tubing and wiring lengths allow for the towers to be mounted approximately 35 meters from the instrument tent. All data are recorded at 20 Hz using a PC-based system designed specifically to record data from the gradient apparatus.

Chamber Method: The flux chamber, which may be operated in dynamic or static mode, is constructed of plexiglass and lined with FEP Teflon™ film. The chamber dimensions are 0.745 m by 0.4 m by 0.145 m yielding a footprint area of 0.298 m^2 and a volume of 0.043 m^3. The method by which the chamber is sealed to the surface of interest varies with the surface conditions. Typically the chamber is mounted on a Teflon™-lined plexiglass corral. The seal with the soil surface is obtained by insertion of the corral to a depth of approximately 4 cm. The chamber design presents a low profile to the wind in order to minimize atmospheric pumping effects. The air inside the chamber is stirred using two Teflon™ fans located in the ceiling of the chamber near the inlet and outlet ports. Use of the fans ensures that the enclosure acts as a continuous stirred tank reactor. The inlet and outlet air concentrations are monitored continuously using two Tekran mercury analyzers. Auxiliary ambient air flow is pulled through the chamber using a vacuum pump to reduce the hydraulic residence time to approximately 4 minutes. The auxiliary flow rate of 10 to 12 L min^{-1} was experimentally determined to minimise the effect of artificially induced flux created by large pressure differentials between the enclosure and the outside atmosphere.

The chamber flux (F) is calculated by the relationship:

$$F = ((C_{out} - C_{in}) * Q)/A,$$

where A is the footprint area (m^2), Q is the flow rate (m^3/s), and C_{out} and C_{in} refer to the outlet and inlet concentrations respectively (ng/m^3), as measured by the Tekran mercury analyzer. Additional supporting measurements for the chamber system include: chamber air temperature and humidity, soil temperature outside the chamber, wind speed and direction, and net radiation.

RESULTS AND DISCUSSION

Natural mercury fluxes have been monitored now at five contrasting geochemical settings (Figure 2). Preliminary results indicate a strong positive relationship (over 3 orders of magnitude) between total mercury concentration in the substrate and the flux of Hg^0 to the atmosphere (Figure 2). The goal of future work is to determine if this relationship holds for a larger data set. If so, it will be possible to use existing geochemical maps to better constrain the natural emissions estimate for Canada. Our future plans are to measure natural Hg^0 emissions from the mercuriferous Pinchi fault zone in British Columbia, from black shales in the Arctic, and from bituminous glacial drift in Saskatchewan.

The average flux of 1,928 ng/m²/hr (1.9 μg/m²/hr) observed in July 1996 at the Clyde Forks sulphide anomaly (Figure 2) is in agreement with earlier flux data cited by the *National Inventory* which range from 0.2 to 1.7 μg/m²/hr (EPS, 1981 citing Lindberg, 1979 and references therein). These observations suggest that the *National Inventory* was correct in using higher emission factors for soils overlying sulphide mineralization compared to other soils. In order to extrapolate over large mineralized areas the *National Inventory* calculated a 2-fold to 4-fold correction factor. This correction factor initially appears conservative in light of the nearly 100-fold difference between average fluxes at the Clyde Forks anomaly site (1,928 ng/m²/hr) compared to the background site (23 ng/m²/hr). The conservative correction factor is more appropriate, however, as the *National Inventory* uses it to calculate average fluxes over extensive regions. Their calculation leads to an estimated natural Hg^0 flux of 32 to 64 ng/m²/hr for soils within British Columbia's mercuriferous belt (EPS, 1981).

Elemental mercury emissions from the Thunder Bay black shale site (Figure 2) were monitored continuously over an 18 day period, averaging 35 ng/m²/hr with daytime highs in July, 1997 of about 100 to 200 ng/m²/hr. The observed fluxes support the practice introduced by the *National Inventory* of using higher emission factors for soils derived from Proterozoic black shales. Although these are the first reported measurements for Hg^0 emissions from this rock type, the results are in agreement with the values predicted by the *National Inventory* (their correction factors result in an estimated flux of 16 to 32 ng/m²/hr). Emissions from black shales of other compositions and locations need to be monitored to determine if the predicted values are adequately representative of the range and variability of fluxes associated with those rocks across Canada.

Large (order of magnitude) temporal variations in Hg^0 flux were observed at each site, suggesting that seasonal variations must be quantified before any attempt is made to calculate a new national emissions estimate. These temporal variations were observed at every site despite the contrasts in mercury speciation and other geochemical characteristics of the substrate from site to site. Daytime averages were typically 4 to 10 times higher than nighttime averages, explained by a strong correlation between Hg^0 flux, air temperature, and solar radiation intensity. Similarly large temporal variations occurred in relation to rainfall events: during a rain event Hg^0 emissions tended to be suppressed, but from the time the rain stopped until the soil dried out, marked increases in fluxes occurred. The latter observation indicates that future measurement campaigns should be undertaken in the spring to determine the variation in Hg^0 fluxes after snowmelt.

Although our data indicate that the *National Inventory* approach resulted in reasonable Hg^0 flux estimates for substrates containing high mercury concentrations (early Proterozoic black shales and sulphides), more flux measurements are required on substrates containing low mercury concentrations (<200 ppb) in order to evaluate the *National Inventory* estimated baseline of 8 to 16 ng/m²/hr. In fact, most of the Canadian landscape is covered with thick deposits of glacial sediments, including unsorted tills, clay, silt and sand deposits. Thus, in order to obtain data that is more representative of the Canadian landscape, a systematic study of Hg^0 emissions from various types of glacial deposits is required.

Sedimentary and sulphide deposits of other geological ages also need to be investigated, as Cameron and Jonasson (1972) and Jonasson and Sangster (1974) have demonstrated that sulphide deposits and black shales of Proterozoic age tend to have higher mercury contents than their Phanerozoic and Archean counterparts.

There are some serious limitations associated with the *National Inventory* approach: that is, with the use of generalized emission factors to quantify the contributions from natural sources compared to anthropogenic sources. It needs to be recognized that the application of a single emission factor to a very large region (such as an entire province) oversimplifies the spatial variability arising from geological influences. For example, regional survey data for glacial deposits covering Precambrian Shield areas of southeastern Ontario and Quebec indicate that natural mercury concentrations within the survey area range from 0.002-1.53 μg/g with a median of 0.060 μg/g and mean of 0.094 μg/g (n = 1, 782 sites), based on analyses of the clay fraction (<2 micron) of samples collected below the weathering zone (Kettles and Shilts, 1994). Once enough data are collected to confirm the relationship shown in Figure 2, it will be possible to use these concentration data to provide an improved natural Hg^0 flux estimate for the area covered by the regional survey. It would be practically impossible, however, to account for small localized anomalies, such as the Clyde Forks anomaly, using the generalized inventory approach. Soils overlying the mineralized fault zone at Clyde Forks, which is located within the survey area, contain natural Hg concentrations as high as 1500 μg/g in the B horizon (4 orders of magnitude greater than the median value), and this small fault zone (100 m in length) alone has probably emitted about 500 kg of mercury to the atmosphere since deglaciation (based on the data presented in Figure 2). Oversimplifications, which are an inherent drawback of the inventory approach, can also result in an overestimate of the natural contribution in a given area. Thus the inventory approach may be potentially misleading for risk assessors and risk managers unfamiliar with geological complexity.

SUMMARY AND CONCLUSIONS

Average flux data acquired in the first two seasons of our field study, based on continuous monitoring for periods of 3.5 to 18 days at each of five contrasting geochemical settings, show a strong positive relationship (over three orders of magnitude) between Hg^0 flux and total Hg concentrations in the substrate. The goal of future work is to determine if this relationship holds for a larger dataset. If so, it will be possible to use existing geochemical maps to better constrain the natural emissions estimate for Canada. Our flux data indicate that the existing *National Inventory* calculated reasonable Hg^0 flux estimates for substrates containing high mercury concentrations due to the presence of early Proterozoic black shales and sulphides. Substantially more flux data are required, however, before any attempt is made to calculate a new estimate for Canada. Limitations of our preliminary data set include the lack of seasonally representative emissions data and the lack of emissions data for substrates having low Hg concentrations.

ACKNOWLEDGEMENTS

The authors thank Tanya Bradley, Laurie Halfpenny, Melody Myers, Keith Neil, Pierre Pelchat, Sandy Steffen, Gabe Eiras and all other staff at Tekran Inc. for their valuable assistance, and Bob Garrett for his careful review of the manuscript. The authors appreciate background information provided by Quentin Bristow, Eion Cameron, Chris Gleeson, Mae Gustin, Ian Jonasson, Don Sangster, and Al Sangster.

REFERENCES

BRISTOW, Q., 1972. Evaluation of the quartz crystal microbalance as a mercury vapour sensor for soil gases. J. Geochem. Explor., 1: 55-76.

CAMERON, E.M. AND JONASSON, I.R., 1972. Mercury in Precambrian Shales of the Canadian Shield: Geochimica et Cosmochimica Acta, 36: 985-1005.

EDWARDS, G.C., SIMPSON, I.J., THURTELL, G.W., KIDD, G.E., 1995. Measurements of the gradients and fluxes of methane and nitrous oxide within and above the forest canopy at the BOREAS Southern Site: Old Aspen Tower. First Science Conference organized with XX General Assembly of the European Geophysical Society, BAHC- Biospheric Aspects of the Hydrological cycle, Hamburg, Germany, April 3-7.

ENVIRONMENT CANADA, 1997. Canadian Sediment Quality Guidelines for Mercury: Pre-publication draft prepared by Guidelines Division, Environmental Conservation, Environment Canada, 102 p. + app.

EPS, 1981. National Inventory of Natural Sources and Emissions of Mercury Compounds; Report by Environmental Applications Group, Ltd. under contract no. 0SS79-000086; Environment Canada.

FRISKE, P.W.B. AND COKER, W.B., 1995. The Importance of Geological Controls on the Natural Distribution of Mercury in Lake and Stream Sediments across Canada. Water, Air and Soil Pollution, 80: 1047-1051.

GEUL, J.J.C., 1970. Geology of Devon and Pardee Townships and the Stuart Location, District of Thunder Bay: Ontario Department of Mines, Geological Report 87, 52p.

HARRIS, D.C., 1989. The mineralogy and geochemistry of the Hemlo gold deposit, Ontario. Geological Survey of Canada, Economic Geology Report 38, 88p.

JONASSON, I.R.,1976. Detailed hydrogeochemistry of two small lakes in the Grenville Geological Province: Geological Survey of Canada Paper 76-13, 36 p.

JONASSON, I.R. AND SANGSTER, D.F., 1974. Variations in the mercury content of sphalerite from some Canadian sulphide deposits, *in* Elliot, I.L. and Fletcher, W.K., eds., Geochemical Exploration: Proc. 5th International Symposium on Geochemical Exploration, Vancouver, Elsevier, Amsterdam, p. 313-332.

KETTLES I. M. AND SHILTS W. W., 1994. Composition of glacial sediments in Canadian Shield terrane, southeastern Ontario and southwestern Quebec: Applications to acid rain research and mineral exploration: Geological Survey of Canada Bulletin 463, 58p.

LINDBERG, S.E., JACKSON, D.R., HUCKABEE, J.W., JANZEN, S.A., LEVIN, M.J. AND LUND, J.R., 1979.Atmospheric Emissions and Plant Uptake of Mercury from Agricultural Soils near the Almaden Mercury Mine. J.Environ. Qual., 8:572-578.

NIKOLS, C.A., 1972. Geology of the Clyde Forks Mercury-Antimony-Copper Deposit and Surrounding Area. Lanark County, Ontario; MSc thesis, Queen's University, Kingston, Ontario.

PAUK, LIBA, 1989. Geology of the Lavant Area, Lanark and Frontenac Counties; Ontario Geological Survey Report 253, 61p. Accompanied by Map 2515, Scale 1:31,680.

RASMUSSEN, P.E., 1994. Current methods of estimating atmospheric mercury fluxes in remote areas. Environmental Science and Technology, 28(13): 2233-2241.

RASMUSSEN, P.E., FRISKE, P.W.B., AZZARIA, L.M., AND GARRETT, R.G.,1998. Mercury in the Canadian Environment. Geoscience Canada, in press.

SCHROEDER, W.H., 1996. Environmental measurements and results of mercury volatilization rates ("fluxes") from soil and lake surfaces. Proceedings of Canadian Mercury Network Symposium; Toronto, Ontario; September 29-30, 1995, EMAN Occasional Paper Series, Report No.6, Ecological Monitoring Coordinating Office, pp. 22-27.

SCHROEDER, W.H., MUNTHE, J., AND LINDQVIST, O., 1989. Cycling of mercury between water, air, and soil compartments of the environment. Water, Air and Soil Pollution, v. 48, pp. 337-347.

VOLDNER, E.C. AND ELLENTON, G., 1987. Production, usage and atmospheric emissions of priority toxic chemicals with emphasis on North America. Prepared for the International Joint Commission. Report: ARD-88-4. Atmospheric Environment Service, 4905 Dufferin Street, Downsview, Ontario, Canada, M3H 5T4.

VOLDNER, E.C. AND SMITH, L., 1989. Production, usage and atmospheric emissions of 14 priority toxic chemicals. International Air Quality Advisory Board of the International Joint Commission, Proceedings of the Workshop on Great Lakes Atmospheric Deposition. Appendix 2.

The European emission inventory of heavy metals for 1990

Jan J.M. Berdowski, J.J. Bloos and A.J.H.Visschedijk
TNO Institute of Environmental Sciences, Energy Research and Process Innovation,
P.O. Box 342
7300 AH Apeldoorn, The Netherlands

ABSTRACT

Within the framework of the OSPARCOM, HELCOM and the UN-ECE a 1990 European emission inventory on heavy metals and persistent organic pollutants has been carried out by the Netherlands Organization for Applied Scientific Research (TNO). The inventory includes all countries in Europe, except the three Caucasian countries and Turkey, and is based on official country submissions and supplementary default emission estimates. Also, within the framework of the UN-ECE Convention on Long-range Transboundary Air Pollution a Heavy Metals emission reduction Protocol is under preparation to reduce air emissions of cadmium, lead and mercury. For this process, additional information on the effects of implementation of the protocol in terms of reduced emissions and incremental costs has been estimated as well. The projections for the year 2010 have been based on the outcome of the 1990 European emission inventory and on RAINS 7.2 scenario assumptions for activity data.

This paper reports on the emission inventory results for 1990 for cadmium, lead and mercury. It presents data for the European countries and for the source categories. Also, it presents the outcome of three different emission scenario's for 2010 for the three heavy metals and the incremental costs of the implementation of the draft Heavy Metals Protocol.

Proceedings of
Metals and the Environment: An International Symposium
Edited by Jim Skeaff
The Metallurgical Society of CIM
100th CIM Annual General Meeting, 1998

INTRODUCTION

At the Third International North Sea Conference it has been decided by Ministerial Declaration that emissions of Priority Hazardous Substances were to be reduced by 1995 (or 1999 at the latest) by 50% or 70% (depending on compound) of the emission level of 1985. The need for emission inventory data to enable monitoring emission reduction processes became more and more clear. The Oslo and Paris Commissions (OSPARCOM) have been requested to co-operate with other relevant international organizations to realize an inventory of atmospheric emissions. These organizations are the Helsinki Commission (HELCOM) and the Executive Body (EB) for the United Nations Economic Commission for Europe (UN-ECE) Convention on Long-range Transboundary Air Pollution (LRTAP). This emission inventory intends to fulfil the requirements for the emission inventories as agreed upon in the work programmes of HELCOM and OSPARCOM and for the development of protocols by the Preparatory Working Groups on Heavy Metals and Persistent Organic Pollutants of the UN-ECE.

Within the framework of the OSPARCOM, HELCOM and the UN-ECE a 1990 European emission inventory on heavy metals and persistent organic pollutants has been carried out by the Netherlands Organization for Applied Scientific Research (TNO) on behalf of the German Environmental Agency (UBA) (Berdowski et al., 1997). The inventory includes all countries in Europe, except the three Caucasian countries and Turkey, and is based on official country submissions and supplementary default emission estimates by TNO. The heavy metals covered by this inventory are arsenic, cadmium, chromium, copper, lead, mercury, nickel, and zinc.

Within the framework of the UN-ECE Convention on LRTAP an emission reduction protocol is under preparation to reduce air emissions of cadmium, lead and mercury. For this process, additional information on the effects of implementation of the protocol in terms of reduced emissions and incremental costs has been considered beneficial. Therefore, the potential costs of the implementation of the protocol and the benefits to future emissions have been determined as well (Berdowski, Pulles and Visschedijk, 1998). These projections for the year 2010 have been based on the outcome of the 1990 European emission inventory (Berdowski et al., 1997) and on RAINS 7.2 scenario assumptions for activity data (IIASA, 1997).

This paper presents the results of the 1990 inventory for cadmium, lead and mercury. The results are a mixture of official country submitted data and default emission estimates by TNO as not for all countries, source categories and substances officials have been in the position to offer complete information. The results for the 2010 emission projections based on RAINS 7.2 scenario assumptions and the incremental costs and benefits of implementation of the HM Protocol are presented as well.

METHODOLOGY

The procedure to prepare the emission inventory has been described by Berdowski et al. (1997). Either country experts submitted emission data for the 11 CORINAIR90 sectors on request of the secretariats of OSPARCOM, HELCOM and UN-ECE or default emission estimates have been set for lacking sectors or countries. In general terms, the emission per source category has been estimated for each country by multiplying the activity rate of the source (e.g. tonnes cement produced) with an emission factor (e.g. tonnes mercury emitted per tonne cement produced). A large number of handbooks on emission factors (Van der Most and Veldt, 1993; Berdowski et al., 1995; McInnes, 1996; Berdowski et al., 1997) and official statistics (see Berdowski et al. 1997) has been used to determine activity rate data and the emissions for the project specific source sectors applied in this inventory.

The following source categories and their sub-sectors have been distinguished for the inventory:

- Public power, co-generation and district heating;
- Commercial, institutional and residential combustion;
- Industrial combustion ;
- Petroleum industries;
- Iron and steel industry: coke production; blast furnace; pig iron; open hearth furnace; basic oxygen furnace; electric arc furnace; rolling mills; sinter plants and foundries;
- Non-ferrous metal industry: aluminium, copper, lead, nickel and zinc industry, and other non-ferrous metal industry;
- Inorganic chemical industry: chloro-alkali industry and NPK fertiliser;
- Organic chemical industry: halogenated hydrocarbons and pesticide production;
- Other industry: cement, glass industry, paper pulp (Kraft process), paper and pulp industry and road paving with asphalt, battery production, synthetic grit;
- Extraction and distribution of fossil fuels;
- Solvent use: paint, industrial degreasing, dry cleaning, wood preservation and other solvent use ;
- Road transport;
- Other transport;
- Waste treatment and disposal: waste incineration, cremation, landfill and mercury thermometers and vapour lamps;
- Agriculture;
- Nature.

The stationary combustion sectors have been the subdivided into consumption data for brown coal, hard coal, fuel oil and other fuels. For the distribution of national country emission totals over the EMEP 50 x 50 km^2 grid both point source information, population density and land use data have been used. These results are not reported in this paper.

The methods applied for emission projections and cost calculation have been described by Berdowski, Pulles and Visschedijk (1998). The activity level projections for 2010 have been estimated according to the Official Energy Pathway (OEP) scenario from RAINS 7.2 (IIASA, 1997).

Three types of emission projections have been prepared for 2010, based on three sets of emission factors:

- applying the OEP Scenario, assuming 'Business as in 1990', without any emission reduction activity and emission factors equal to those for 1990;
- applying the OEP Scenario, assuming a range of autonomous emission reducing activities (i.e. non-protocol related measures, both country and region specific: elimination of leaded gasoline, the EURO-CHLOR program for the chloro-alkali-industry, national emission regulations, European Union Directives for EU-countries, implementation of the Second Sulphur Protocol); these measures will influence the emission factors;
- applying the OEP Scenario, assuming autonomous emission reducing developments and implementation of the Protocol measures for heavy metals emitting stationary sources and heavy metals containing products (UNECE, 1997); also these measures will influence emission factors.

Both investment costs and the resulting annual costs have been estimated. The annuity has been calculated by using (Kok, 1997):

$$Annuity = \frac{i(i+1)^n}{(1+i)^n - 1}$$

with i = Interest rate (-)
n = Amortisation period (years)

Based on the technical life expectancy of the emission control units, an amortisation period of 20 years has been chosen. The interest rate has been fixed on 4%.

The annuity of the total capital investment (TCI) has been increased by the operational costs in order to come to the total annual costs. The operational costs are estimated by taking a percentage of the TCI (5%).

The cost estimates have been made according to two different starting points:

- the costs required to come to the emission levels after implementation of the Protocol in absence of any other emission reducing activity;
- the costs required to implement the Protocol after all other autonomous emission reducing activities have been implemented.

RESULTS

Table 1 presents the total European emissions of cadmium, lead and mercury in 1990. These are 612, 40,400 and 463 tonnes respectively. Also, the data per country are presented in the Table 1.

It has to be kept in mind that the data used are a mixture of information submitted by a number of countries and default estimates by TNO. About 55-65% of the total emissions presented in Table 1 originate from official countries' submissions. The data submitted for this inventory have not been in a form enabling uncertainty analysis, but the uncertainty in the emissions will be large. However, on the basis of a validation study on the atmospheric deposition of heavy metals on the North Sea (Baart, Berdowski and Van Jaarsveld, 1995), using comparable emission data, the overall uncertainty for the northwestern European countries is estimated to be 20 to 50%. For central and eastern European countries the uncertainty will be larger and the maximum margin of the uncertainty range might well be 2.5-3.5.

TABLE 1 Annual Emissions of Cadmium, Lead and Mercury in Europe in 1990 by country (tonnes/year).

Country	Cadmium	Lead	Mercury
Albania	0.647	33.4	0.511
Austria	5.11	215	4.27
Belgium	9.9	716	8.86
Bulgaria	8.41	317	6.9
Bosnia and Herzegovina	0.404	8.63	0.208
Belarus	6.58	736	0.091
Switzerland	4.24	520	6.82
Cyprus	0.197	0.906	0.305
Czech Republic	12	337	9.31
Germany	31.5	2347	113
Denmark	2.12	179	6.92
Spain	36.7	4674	20.2
Estonia	3.87	171	2.03
Finland	3.66	215	3.04
France	14.8	4414	32.5
United Kingdom	24.9	2703	25.6
Greece	4.47	505	7.12
Croatia	3.24	466	1.08
Hungary	4.6	639	4.2
Ireland	1.59	134	1.62
Iceland	0.166	6.38	0.048
Italy	59.8	1642	11.8
Lithuania	2.84	246	0.003
Luxembourg	1.14	73.5	0.767
Latvia	3.23	218	0.336
Moldova	1.78	168	1.52
Former Yugoslav Republic of Macedonia	9.13	210	1.49
The Netherlands	2.17	266	2.64
Norway	2.42	226	2.34
Poland	91.6	1372	33.3
Portugal	2.98	631	5.48
Romania	21.6	585	7.5
Russian Federation	159	10148	86.2
Slovak Republic	9.66	166	12.4
Slovenia	1.03	123	0.869
Sweden	2.04	537	1.45
Ukraine	54.2	3878	36
Yugoslavia (Serbia and Montenegro)	8.3	597	3.86
Total Europe	**612**	**40400**	**463**

Figure 1 presents an overview of the relevance of source categories for the emissions of cadmium, lead and mercury in Europe in 1990.

Figure 1 Relative contribution of source categories to total annual emissions of cadmium, lead and mercury in Europe in 1990.

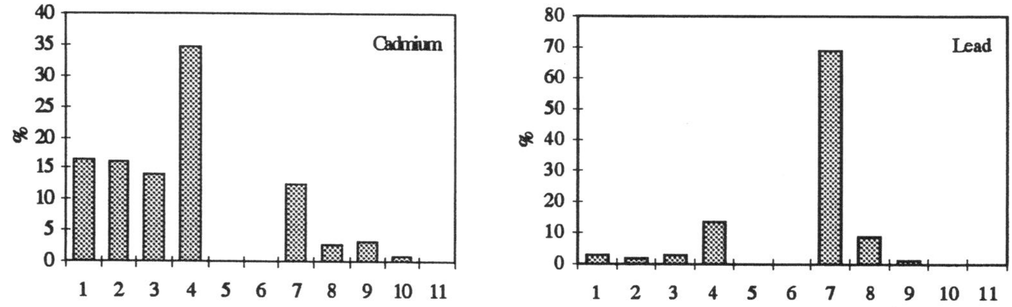

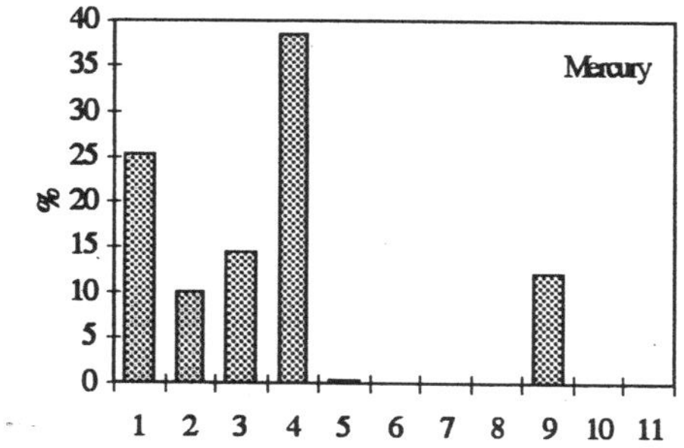

1 Public power, cogeneration & district heating
2 Commercial, institutional & residential combustion
3 Industrial combustion
4 Production processes
5 Extraction & distribution of fossil fuels
6 Solvent use
7 Road transport
8 Other mobile sources & machinery
9 Waste treatment & disposal
10 Agriculture
11 Nature

The major contributing source categories to total European emissions are:

- for cadmium, the emission is mainly caused by combustion of hard and brown coal and fuel oils, road transport and production processes in the zinc and the iron and steel industry;
- for lead , the emission is mainly caused by combustion in all transport and production processes;
- for mercury, the emission is mainly caused by hard and brown coal combustion, production processes in the cement and non-ferrous metal industry and municipal waste incineration.

The results of the emission projections are presented in Figure 2.

Figure 2 Total cadmium, lead and mercury emission in Europe in 1990 and three scenario forecasts for 2010

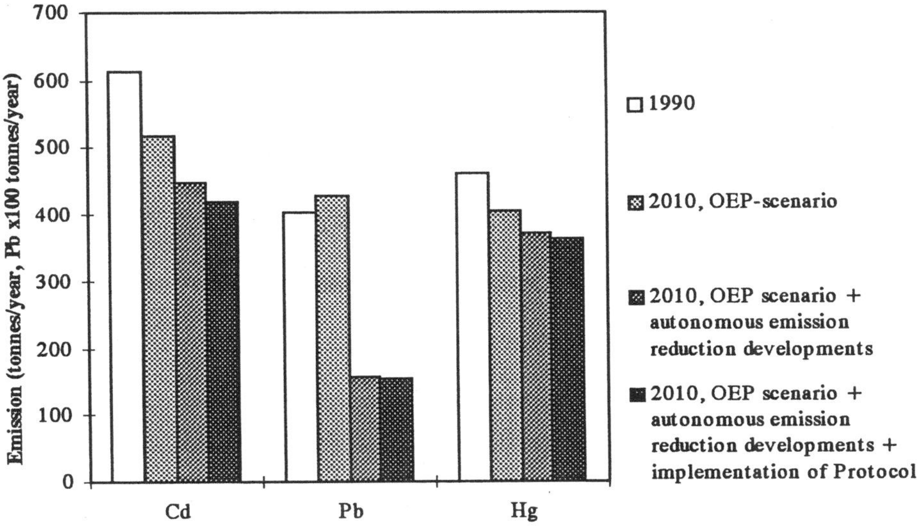

For cadmium and mercury, a clear decrease in emissions is expected applying the assumptions of the Official Energy Pathway (OEP) scenario, both without any emission reducing activity and under the assumption of a range of non-Protocol related autonomous emission reducing activities. For lead emissions, an increase has been forecast in the case of absence of autonomous emission reducing activities as the OEP scenario does not count for further penetration of unleaded gasoline in traffic. When this development is taken into account a emission reduction of about 60% is expected. For all three heavy metals, the incremental emission reduction of the implementation of the Protocol is relatively limited. As a result of both autonomous emission reducing activities and the implementation of the Heavy Metals Protocol the emissions of cadmium, lead and mercury will decrease with 21-62% (Table 2).

TABLE 2 Forecast Emission Reduction of Cadmium, Lead and Mercury in Europe after the Implementation of the Heavy Metals Protocol.

Substance	Emission level in 1990 (tonnes)	Emission level in 2010 after Protocol (tonnes)	Reduction (%)
Cadmium	610	420	32
Lead	40,000	15,000	62
Mercury	460	360	21

Remaining large emission sources after the implementation of the Heavy Metals Protocol will be:

- for cadmium, residential combustion in Central and Eastern Europe, road transport and the zinc production industry (although the emission data for the zinc industry are rather uncertain);
- for lead, remaining emissions from the use of gasoline (when still containing some lead);
- for mercury, coal firing in general and the cement production industry (mercury removal from large waste gas flows from these sources is difficult to conduct).

The total annual costs of implementation of measures listed in the UN/ECE draft HM Protocol are estimated to amount to 440 MECU/year when also the foreseen autonomous emission reducing developments would take place. The costs would amount to 1300 MECU/year without any of such autonomous developments. By far the largest contribution comes from emission control at stationary sources (Figure 3).

Figure 3 Annual incremental cost of the Heavy Metals Protocol (440 MECU/year) by sector in Europe after autonomous developments.

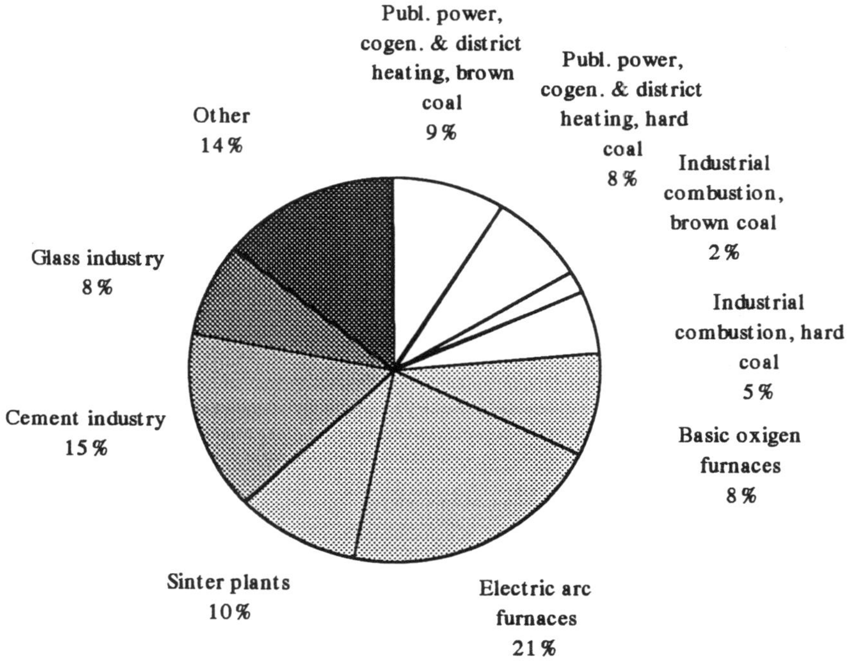

In Figure 3, the incremental costs of the HM Protocol implementation has been indicated and the distribution over the sectors after foreseen autonomous measures have been taken. The total capital investment costs are estimated to be 3,5 billion ECU with and 11 billion ECU without any autonomous emission reducing activities.

Protocol-related emission reduction measures concerning restrictions on or elimination of the use heavy metal containing products are estimated not to significantly influence the total costs of the HM Protocol when expected autonomous developments are counted for.

REFERENCES

BAART, A.C., BERDOWSKI, J.J.M. and VAN JAARSVELD, J.A., 1995.
Calculation of atmospheric deposition of contaminants on the North Sea. TNO-MEP report R 95/138, TNO, Delft, 122 p.

BERDOWSKI, J.J.M, VELDT, C., BAAS, J., BLOOS, J.P.J., and KLEIN, A.E.,1995.
Technical Paper to the OSPARCOM-HELCOM-UNECE Emission Inventory of Heavy Metals and Persistent Organic Pollutants. TNO-MEP report R 95/247, TNO, Delft, 42 p.

BERDOWSKI, J.J.M, BAAS, J., BLOOS, J.P.J., VISSCHEDIJK, A.J.H. and ZANDVELD, P.Y.J.,1997.
The European Inventory of Heavy Metals and Persistent Organic Pollutants for 1990. UBA Forschungsbericht 104 02 672/03B, UBA, Berlin, 239 p.

BERDOWSKI, J.J.M, PULLES, M.P.J. and VISSCHEDIJK, A.J.H., 1998.
Incremental cost and remaining emission in 2010 of Heavy Metals (HM) resulting from the implementation of the draft HM Protocol under the UN/ECE Convention on Long Range Transboundary Air Pollution. TNO-MEP report R 98/020, TNO, Apeldoorn, 69 p.

IIASA, 1997.
Regional Air Pollution Information and Simulation Model, Version 7.2, International Institute for Applied System Analysis (IIASA), Laxenburg.

KOK, H., 1997.
Technical Information from TNO-MPT, Apeldoorn.

MCINNES, G., 1996.
Atmospheric Emission Inventory Guidebook. A Joint EMEP/CORINAIR Production, prepared by the EMEP Task Force on Emission Inventories, Copenhagen.

UN/ECE , 1997.
Technical annex to the HM emission Protocol under preparation, UN/ECE, Geneva.

VAN DER MOST, P.F.J. and VELDT, C., 1993.
Emission factors manual PARCOM-ATMOS. Emission factors for air pollutants 1993. TNO report R 92/235, TNO, Apeldoorn, 64 p.

Trace substances emissions from U.S. electric utilities and their implications

P. Chu, L. Levin and M.A. Allan
Electric Power Research Institute
3412 Hillview Ave.
Palo Alto, California 94303 U.S.A.

ABSTRACT

Since before the 1990 U.S. Clean Air Act Amendments, EPRI has been collecting data on trace substances emissions, including trace metals, from fossil-fueled electric generating power stations. Coupled with an intensive effort to assemble a more credible and consistent dataset through field testing between EPRI and the U.S. Department of Energy is EPRI's work on mercury and nickel species measurement methods development, on EPRI's Mercury Cycling Model, and on human exposure and health risk assessments of selected metals such as arsenic and mercury. In 1994, EPRI published an electric utility power plant hazardous air pollutants assessment synthesis report and several mercury emission addendums. This technical paper will summarize the trace element emission results reported in various EPRI reports and EPRI's continuing emissions, exposure, and risk assessment research to date.

Proceedings of
Metals and the Environment: An International Symposium
Edited by Jim Skeaff
The Metallurgical Society of CIM
100th CIM Annual General Meeting, 1998

INTRODUCTION

Title III of the 1990 Clean Air Act Amendments (CAAA) addresses human exposure to trace substances listed in the legislation as "hazardous air pollutants". This portion of the amendments contains specific provisions that require the U.S. Environmental Protection Agency (EPA) to conduct a study of the potential health risks posed by the emission of hazardous air pollutants from electric utility plants. The legislation explicitly directs EPA to focus on the risks to human health due to emissions from fossil fuel-fired electric utility steam generating units (with >25 MW capacity) *after* utilities have complied with other provisions of the Clean Air Act (such as provisions related to emissions of sulfur and nitrogen oxides). The EPA administrator is to decide whether further regulation is "necessary and appropriate" based on the results of the EPA study.

The Electric Power Research Institute (EPRI) integrated the best scientific data and methods currently available to understand the potential magnitude and nature of human health risks due to trace substances emissions from electric utility steam-generating units in the U.S. EPRI's Trace Substances Synthesis Report summarizes the results of recent research, conducted by EPRI and others, addressing a range of critical issues such as trace substances emission rates from utility generating units, appropriate sampling and analytical methods for conducting flue gas measurements, and the toxicity of specific substances found in utility emissions[1].

The results of the measurement programs were combined with other data to evaluate community health risks from each of 1750 generating units at each of 600 power plants, as projected in the year 2010. The data, methodologies, and analysis results are discussed below.

METHODS

Emissions Measurements

In order to improve the database on emissions from operating utility units, a measurement program organized under the EPRI PISCES program carried over 40 field sampling measurements at coal-, oil-, and gas-fired utility power plants in the U.S. Sites were chosen to encompass every major classification of fuel type, boiler design, and particulate, NO_X, and SO_2 control technologies. This field sampling included trace substance measurements in the flue gas stack and often at the air pollution control device inlet. State-of-the-art sampling and analytical protocols were employed in order to quantify many of the trace substances[2]. The substances measured included:

1. Trace metals including arsenic, beryllium, cadmium, chromium, lead, manganese, mercury, nickel, selenium
2. Anions including HCl, HF
3. Aldehydes including formaldehyde
4. Volatile organic compounds (VOCs) including benzene, toluene
5. Polynuclear aromatic hydrocarbons (PAHs)
6. Dioxins/furans

The fuel, bottom ash, fly ash, and wastewater streams were also analyzed for trace substances in order to conduct material balance calculations. Additional data from field measurements at eight coal-fired power plants tested by the U.S. Department of Energy [3,4] were incorporated into EPRI's data analyses. In addition to these field studies, EPRI and the Utility

Air Regulatory Group (UARG) conducted a study of mercury and trace metals in over 150 as-fired coal samples[5].

Industrywide Emissions Assessment

The results from the EPRI and DOE field sites were used to estimate emissions for all 1750 operating units at the 600 power plants in the U.S. Several approaches were used, and varied depending upon the factors that affect emissions [6,7]. Power plants were subdivided by fuel type (coal, oil, and gas) and control technologies. The methodology varied as well for the various trace substances. For the metals primarily in the particulate phase (such as arsenic, chromium, lead, nickel) in stack emissions at coal-fired power plants, a correlation was developed to estimate emission factors (mass emitted per heat input - $lb/10^{12}$ Btu) with the concentration in the coal, ash fraction of coal, and total particulate emissions rate. Figure 1 illustrates the correlation for arsenic emissions for coal-fired power plants. This correlation combines all data from the various coal ranks, boiler types, and air pollution control technology. While a specific coal type may generally yield higher concentrations of select trace metals, this is accounted for in the "concentration in the coal" variable. Likewise, control technology performance is incorporated in the "particulate emissions" variable. Similar correlations were developed for all the particulate phase metals.

For plants burning fuel oil and natural gas, an average emission factor was most appropriate for the particulate-phase metals. This was done, in part, due to the lack of a database to estimate concentrations of trace metals in the fuel oil and natural gas. Similarly, all other trace substances for oil- and gas-fired plants were estimated using average emission factors.

For more volatile trace elements (mercury, selenium, HCl, HF), average control efficiencies were applied to the elemental content of the coal. Mercury removal efficiencies were highly variable, however a predictive model could not be developed. Mercury emissions are believed to be strongly a function of the form or speciation of mercury [8]. Current research by EPRI, DOE, and others are attempting to predict mercury speciation and control technology performance. For organic emissions which are believed to be produced during the combustion process, an average emission factor was applied for each organic compound.

These parameterizations were applied to every US power plant operating in 1990. Emission estimates for an hypothetical 500 MW bituminous coal-fired power plant with an ESP is summarized in Figure 2. Annual trace metal emissions for this hypothetical power plant tend to be about or less than 100 kg/year. Emissions of HCl and HF are several orders of magnitude greater and are not included in Figure 2. In addition, these compounds are not carcinogenic, and these emissions were also found to have negligible non-cancer risks. The estimated total US utility emissions in 1990 for various trace metals are summarized in Figure 3. Figure 3 also includes projected emissions in 2010, as defined in the Clean Air Act Amendments.

Emissions estimates for mercury have been developed for the United States in recent years, as shown in Table 1. Total emissions have changed significantly from the first estimate published by U.S. EPA in 1993 (301 Mg/yr) and that most recently reported in the U.S. EPA, 1997 (144 Mg/yr). Looking at Table 1, estimates appear to be a decreasing over time, with the recent totals about 50% less than the first reported. However, this apparent trend is due in large part to improvements in the methods and data used to estimate mercury emissions, and in lesser part to actual reductions in emissions. The earlier inventories were for the base year 1990, with newer information incorporated into the more recent inventories, which were finally converted to a 1994-1995 base year in the most recent U.S. EPA report. Certain categories may have experienced reductions due to changes in mercury use, disposal and regulatory control, most notably medical and municipal waste incineration. These inventories should by no means be viewed as static: improved measurement methods will continue to yield better estimates for these source categories and other categories that are not included in the current inventories due to lack of data. It should also be noted that the inventories are for anthropogenic sources only, and do not include natural emissions from the terrestrial landscape, which may be a sizable source of gaseous mercury to the atmosphere.

Exposure and Health Effects

The estimated trace substanc7e stack emissions were used as input to the EPA ISCLT2 atmospheric dispersion model in order to calculate downwind ground-level concentrations. These concentrations, particularly the maximum ground-level concentrations, were used in combination with exposure scenarios to represent long-term inhalation exposure concentrations, or exposure doses. The exposure dose information for each substance was then combined with health effects information on the respective substances to calculate inhalation health response and consequent risks.

Risk Assessment

For carcinogenic effects, risk is measured as the probability of a cancer occurrence over a 70-year lifetime. Risks were calculated for the "maximally-exposed individual (MEI)", where all assumptions of exposure to outdoor air concentrations are at their most conservative, as well as for the "reasonably-exposed individual (REI)", which assumed a more "realistic" pattern of human activity throughout a lifetime of exposure. Figure 4 displays the distribution for all power plants of the inhalation carcinogenic risks for the year 2010 (using the "EPA" 2010 scenario for SO_2 control), for both the MEI and REI exposure scenario. Even for the "worst-case" MEI assumption, only 3 power plants (out of approximately 600) display inhalation carcinogenic risks above one in one million (1×10^{-6}) - a level below which regulators will commonly not require more detailed studies. For REI assumption, the risks for all plants fell well below one in one million.

Since mercury (and other substances) have pathways to human exposure through environmental media other than air, a more complex set of exposure models was developed to provide insight into these alternative pathways. This model, TRUE (Total Risk of Utility Emissions), was applied to 4 case study power plants for which measurement and environmental data were available. Table 2 summarize the multimedia risk results for carcinogenic risks for the four case studies. The MEI carcinogenic risks for each case study were much less than 1×10^{-6}.

Control Technologies

EPRI is currently sponsoring research to evaluate and develop various technologies to reduce trace metal emissions. The primary focus has been on Hg, because mercury appears to be the trace element of highest concern. Both pre- and post-combustion technologies are being evaluated, including coal cleaning, activated carbon injection, and flue gas desulfurization technologies.

CONCLUSIONS

EPRI conducted an extensive field measurement program that provides a database for predicting trace substances emissions for all 600 power plants. The emissions results were integrated along with other research in fate and health effects into a nationwide assessment of health risks. Even when conservative estimates of exposure are used, only 3 out of 600 power plants exhibit inhalation carcinogenic risks which might require more detailed site-specific study (greater than 1 in a million). When more reasonable estimates of exposure, the health risks drop by 50 to 80 percent - with no plant having risks greater than 1 in a million. Four case study evaluations of multimedia risk were conducted and showed risk levels below currently-listed and proposed federal guidelines.

REFERENCES

1. Electric Utility Trace Substances Synthesis Report, Report TR-104614-V1 through -V4, Electric Power Research Institute, Palo Alto, California, 1994.

2. Field Chemical Emissions Monitoring (FCEM) Generic Sampling and Analytical Plan, Report TR-104869, Electric Power Research Institute, Palo Alto, California, 1995.

3. Schmidt, C.R. and T.D. Brown, "Comprehensive Assessment of Air Toxics Emissions from Coal-Fired Power Plants", Proceedings of the 87th Annual Meeting and Exhibition of the Air & Waste Management Association, Paper 94-WA-68A.03, June 19-24, 1994.

4. A Comprehensive Assessment of Toxic Emissions from Coal-fired Power Plants: Phase 1 Results from the U.S. Department of Energy Study, USDOE, September 1996.

5. Mercury and Other Trace Metals in Coal, Report TR-106950, Electric Power Research Institute, Palo Alto, California, 1997.

6. Behrens, G., Chu, P., Chow, W., and R. Roberson, "A Method for Estimating Emissions of Trace Substances from Power Plants", Proceedings of the 87th Annual Meeting and Exhibition of the Air & Waste Management Association, Paper 94-WP-68B.02, June 19-24, 1994.

7. Emission Factors Handbook: Guidelines for Estimating Trace Substance Emissions from Fossil Fuel Steam Electric Plants, Report TR-105611, Electric Power Research Institute, Palo Alto, California, 1995.

8. Mercury in the Environment - A Research Update, Report TR-107695, Electric Power Research Institute, Palo Alto, California, 1996.

Table 1 Comparison of Estimated Annual Mercury Emissions (in metric tons per year)

Source Category	U.S. EPA, 1993[1,2] ~1990	U.S. EPA, 1996[3] ~1990	Pai, et al., 1998[4] 1990+	U.S. EPA, 1996[5] 1994-1995
Electric Utility Fuel Combustion	94.5	46.5	42.5	47.2
Industrial/Commercial Fuel Combustion	31.4	26.3	18.5	25.8
Residential Fuel Combustion	31.2	3.2	0.8	3.3
Municipal Waste Combustion	59.5	50.0	45.5	26.9
Medical Waste Incineration	58.7	58.8	11.1	14.6
Manufacturing/Smelting	8.2	29.1	26.4	14.4
Miscellaneous	17.7	6.2	1.6	11.9
TOTAL	**301.2**	**220.1**	**146.4**	**144.1**

[1]U.S. Environmental Protection Agency, 1993. Locating and Estimating Air Emissions From Sources of Mercury and Mercury Compounds. EPA-454/R-93-023. Midwest Research Institute.

[2]Electric Power Research Institute. 1994. Electric Utility Trace Substances Synthesis Report, Volume 3. TR-104614-V3. Electric Power Research Institute, Palo Alto, CA

[3]U.S. Environmental Protection Agency, 1996. Mercury Study Report to Congress. SAB Review Draft. EPA-452/R-96-001. Office of Air Quality Planning & Standards and Office of Research and Development, Washington DC.

[4]Pai, P., Heisler, S., and Joshi, A., 1998. "An Emissions Inventory for Regional Atmospheric Modeling of Mercury," Water, Air and Soil Pollution (in press).

[5]U.S. Environmental Protection Agency, 1997. Mercury Study Report to Congress, Volume II: An Inventory of Anthrophogenic Mercury Emissions in the United States. EPA-452/R-97-003. Office of Air Quality Planning & Standards and Office of Research and Development, Washington DC.

Table 2: Carcinogenic Risk Results, TRUE Multimedia Case Studies

	Site 1	Site 2	Site 3	Site 4
Dominant Chemicals	Arsenic Chromium (VI) Cadmium	Arsenic Chromium (VI) Cadmium	Arsenic Beryllium	Beryllium
MEI Risk	1.4×10^{-8}	4.0×10^{-8}	6.0×10^{-7}	2.0×10^{-8}

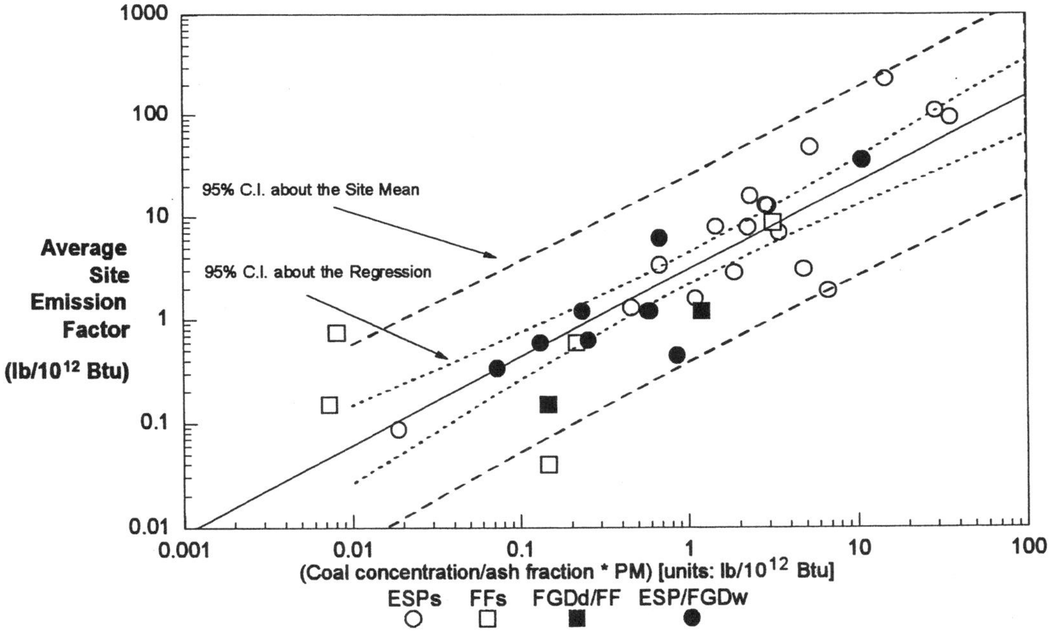

Figure 1. Data points and derived fit for arsenic emissions from bituminous coal-fired power plants.

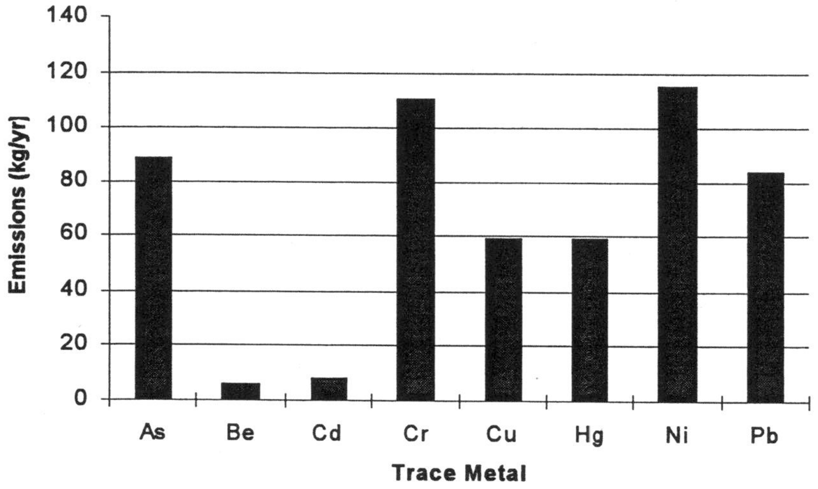

Figure 2. Trace metal emissions from a hypothetical 500 MW bituminous coal-fired power plant, with an ESP.

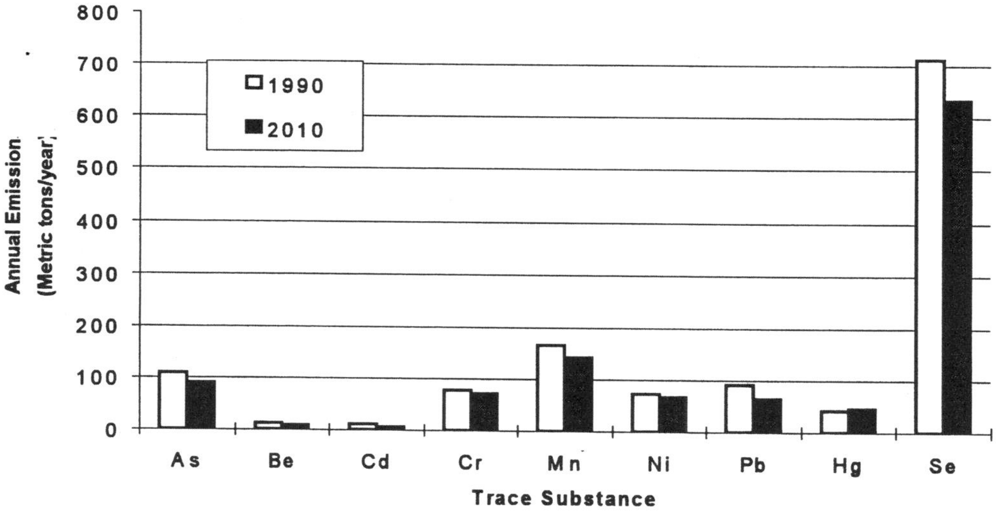

Figure 3. Estimated US Utility Trace Metal Emissions in 1990 and 2010

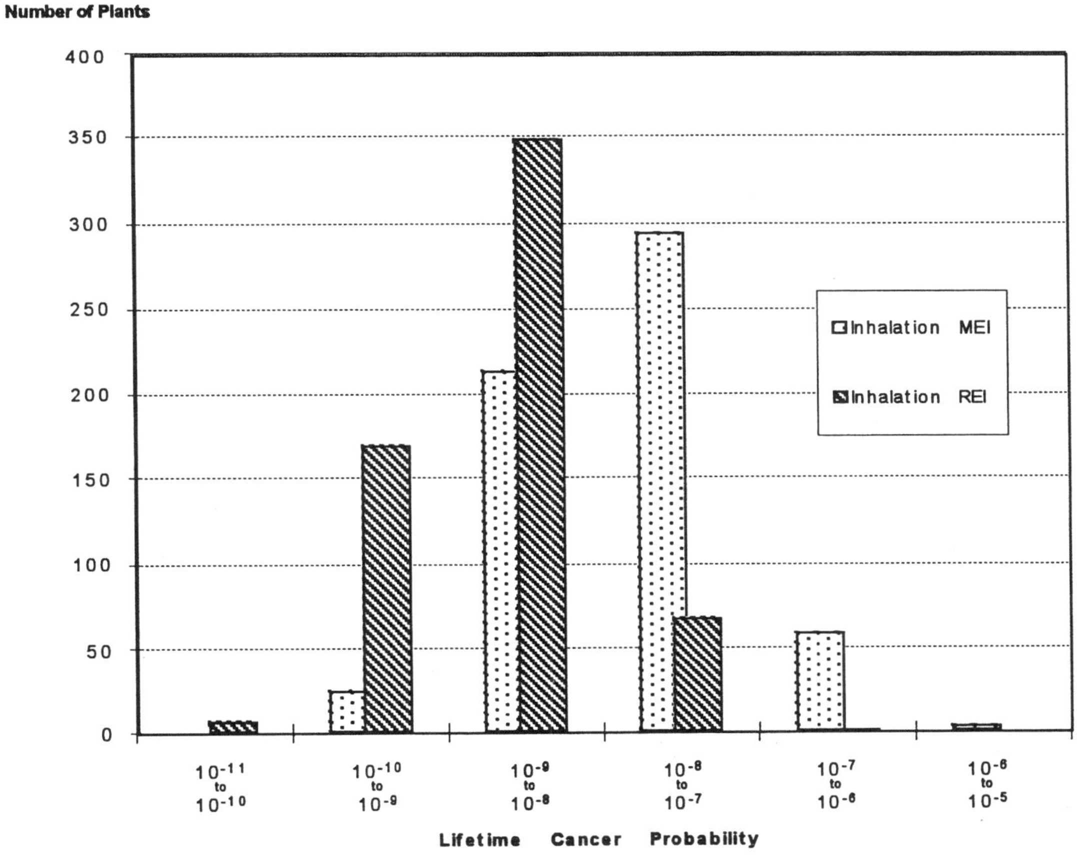

Figure 4. Distribution of inhalation carcinogenic risk, reasonable exposure vs. maximal exposure

Calculated emission factors of trace metals for Canadian non-ferrous smelters [1]

J.M. Skeaff and A.A. Dubreuil
Environment Laboratory, Mining and Mineral Sciences Laboratories, CANMET
Department of Natural Resources, 555 Booth St., Ottawa, Canada K1A 0G1

ABSTRACT

Emission factors of As, Cd, Cu, Hg, Mn, Ni, Pb, Sb, Se and Zn from the Canadian primary non-ferrous smelting industry have been calculated from publicly available 1993 emission release data and metal production data. With a few exceptions, the calculated weighted average 1993 emission factors for Canadian lead, copper+nickel and zinc metal production by non-ferrous smelting are between approximately 40% and more than 1,500 times lower than published emission factors of As, Cd, Cu, Hg, Mn, Ni, Pb, Sb, Se and Zn used in the literature to estimate global inventories of emissions from the world smelting sector. Where comparisons are possible, these calculated emission factors validate, with two exceptions, the EMEP/CORINAIR values suggested to estimate release inventories from lead and copper(+nickel) production; however, use of the EMEP/CORINAIR values for zinc production would considerably overestimate the releases from the primary Canadian zinc industry. With two exceptions, emission factors for As, Cd, Hg, Ni and Pb for lead, copper+nickel and zinc production show incremental to marked decreases between 1993 and 1995.

Key word index: Emission factors, trace metals, heavy metals, non-ferrous smelting, atmospheric emissions.

1 This is an abridged and updated version of *Atmospheric Environment*, 31(10), Skeaff, J.M. and Dubreuil, A.A., "Calculated 1993 Emission Factors of Trace Metals for Canadian Non-Ferrous Smelters", 1449-1457, Copyright 1997, with kind permission from Elsevier Science Ltd., The Boulevard, Langford Lane, Kidlington OX5 1GB, U.K.

INTRODUCTION

Emission factors can be considered as estimates of the amounts of trace elements released per unit produced, all on an annualised basis, and are used in a "Sufficient Weight of Evidence" approach to reckoning anthropogenic trace metal release inventories from various processes to the atmosphere. While inadequate for calculating the absolute quantities of trace elements released during processing, in the absence of valid data, emission factors can be useful as a rough guide to trace metal releases, and are used from time to time in the setting of national and international environmental policies, protection measures and regulations. To ensure that expenditures on pollution control measures are properly targetted and warranted, it is of some importance that emission factors be based on the most recent and accurate data. In this paper, we will use the symbol ε for emission factor, and the units will be g of trace metal emitted per tonne of metal produced (e.g., g As/t Zn).

Nriagu and Pacyna (1988) were apparently the first to use aggregated emission factors to estimate trace metal emission inventories to air, water and soil on a global basis from such anthropogenic sources as coal and oil combustion, non-ferrous metal production, steel and iron manufacturing, and municipal incineration. As their estimates were for 1983, one might anticipate that some detectable improvements in trace metal emission controls may have occurred since then.

However, as far as we are aware, until now little attention has been paid to validating trace metal emission factors from the primary non-ferrous metals producing industry by comparisons with reported emission data. It is of interest to compare aggregated emission factors for trace metals that have been developed for the worldwide non-ferrous metals production industry (Nriagu and Pacyna, 1988; EMEP/CORINAIR 1995) with those that can readily be calculated from publicly available metal production (NRCan, 1994a, 1994b) and from trace metal emission data (BMS, 1994a; EC, 1995, 1997; MAC-ARET, 1995, 1997; MacLatchy, 1992) for Canadian non-ferrous smelters and refiners.

The quantities, identity and characteristics of trace metals emitted from non-ferrous metal production will depend to a great extent on a wide range of factors:

- type of base metal ore from which the smelter concentrate feed is produced - e.g., Pb-Zn, Cu-Ni, Cu-Zn and whether the ore is an oxide or sulphide (essentially, Canadian non-ferrous smelters treat only metal sulphide concentrates);
- ore mineralogy - e.g., relative amounts of constituent sulphide minerals such as covellite, pentlandite, bornite, galena, sphalerite, chalcopyrite;
- character and distribution of trace and minor minerals among bulk minerals - e.g., greenockite (CdS) can be associated with sphalerite (ZnS) but is seldom found with pentlandite (Fe,NiS);
- beneficiation processes - e.g., column flotation, reagents, pH, temperature;
- type of pyrometallurgical process - e.g., Pierce-Smith copper converting, Mitsubishi, Noranda, Inco or Outokumpu flash smelting (Biswas and Davenport, 1980);
- particulate and trace metal emission controls - e.g., sulphuric acid plant, electrostatic precipitators (ESP's), Cottrell precipitators, filter bags.

Obviously, a low emission factor for one trace metal from an individual smelter may not necessarily mean that the smelter's emission controls are performing well, but only that the feed to the smelter contains negligible amounts of the trace metal.

In this report, we present the results of the calculations of emission factors from Canadian non-ferrous smelters for the trace metals Cd, Hg and Pb, which are of significance regarding a proposed protocol for "heavy metals" under the United Nations Economic Commission for Europe (UN-ECE) convention on Long-Range Transboundary Air Pollutants (LRTAP); we have also calculated emission factors for As, Cu, Mn, Ni, Sb, Se and Zn, as well as those of As, Cd, Hg, Ni, and Pb for 1995 which may be of importance in other contexts. We compare our calculated Canadian emission factors with those in the literature that have been used to estimate global inventories of smelting sector emissions. We should finally note that, while for the sake of brevity, trace metal emissions are discussed here as elements, they are in fact seldom if ever emitted from smelters in the metallic form but rather in compounds such as complex oxides, sulphates and silicates (Eatough *et al.*, 1978), part of the mercury in process gases being an exception (Daum and Sneyd, 1992).

METHODOLOGY

Given the wide currency frequently attained by published emission factors, we have previously presented our methods of calculating the trace metal emission factors for Canadian non-ferrous smelters in some detail (Skeaff and Dubreuil, 1997), as the procedures for estimating emission factors are not always adequately addressed in the literature.

We obtained data on trace metal emissions to air as per BMS, 1994a; EC, 1995, 1997; MAC-ARET, 1995, 1997; MacLatchy, 1992. In Canada, the emission of a listed substance must be reported to the National Pollutant Release Inventory (NPRI) if its concentration exceeds 1% in any process stream or if a total of 10 t of the substance is processed in a plant within a calendar year (EC, 1995). We have counted as air emissions both stack and fugitive release data from NPRI, and this, in a few instances, nearly doubles the atmospheric emission from a smelter.

We used both reported smelter capacity figures NRCan (1994a, 1994b) and literature data (references given in Table 1) to calculate realistic estimates of metal production on an individual smelter basis for the early-mid 1990's, taking data for 1993 wherever possible.

In Table 1, we have shown the metal tonnages produced, the reported trace metal emissions, e_C, from Canadian smelters for 1993, and from these the calculated emission factors, ε_C, for the trace metals. The sources of Canadian emission data are noted in Table 1 beside the emission values.

In cases wherein a Canadian smelter produces only one metal, we simply calculated the emission factor by dividing the smelter's reported trace metal emission by the tonnes of metal produced. For BMS, the 1993 NPRI emission value of As is 5.5 t (EC, 1995) and BMS produced about 74,000 t of Pb in 1993, so the BMS emission factor for arsenic in 1993 is:

$$\varepsilon_C = 5.5 \times 10^6/74{,}000 = 74.3 \text{ g As/t Pb} \qquad \text{... eqn. 1}$$

Trace metal emissions are assigned to individual smelters (e.g., EC, 1995; MAC-ARET, 1995) and not broken out into the metals they produce, so if a smelter such as Cominco produces two metals, then in the absence of data indicating that a trace metal was preferentially concentrated more into one process stream than into the other, we prorated the trace metal emissions on the basis of the respective tonnes of metals produced. In the case where a smelter produces two metals, trace amounts of which are both emitted, we estimated the amount of produced metal in each concentrate and prorated the trace metal emissions according to their amounts in the respective concentrates. Finally, we have calculated the weighted average emission factors, ε_{WA}, for each trace metal (Skeaff and Dubreuil, 1997).

Beyond the simple prorating we have done here for smelters that produce more than one metal, it will be difficult to assign trace metal emissions to the different metals produced, since the metal circuits may be interrelated and quite complex. Finally, since a copper smelter may rely, in times of low base metal prices, on precious metal recovery for its continued operation, one could ask whether all or part of the trace metal emissions should be assigned to gold and silver production instead of to copper or zinc (Webster, 1995).

In the next section, we compare the total reported trace metal releases, ε_C, for the Canadian smelting industry with those calculated from the "median" (actually the mid-point) values of the range of emission factors, ε_P, estimated for the non-ferrous smelting industry world-wide and used by Nriagu and Pacyna (1988) to calculate global emission inventories from that sector. However, the median and mid-point of their ranges will not in general coincide unless the distribution is symmetrical. If a distribution is skewed by a few, perhaps small in capacity, smelters that have very high emission factors, the mid-point of the range could give higher estimates than appropriate to represent smelter emissions on a global basis (Webster, 1995).

DISCUSSION

As might be expected, the calculated Canadian trace metal emission factors for 1993 given in Table 1 for As, Cd, Cu, Hg, Mn, Ni, Pb, Sb, Se and Zn show significant variations on a smelter-by-smelter basis, from 0.0 g As/t Zn for CEZ to about 11,500 g Zn/t Zn for HBMS. (With the closing of its zinc concentrate roasting plant in October 1993 (Collins *et al.*, 1994), HBMS no longer emits trace metal-bearing particulates to the atmosphere from its zinc operations.) On an industry-wide basis, the weighted average emission factors vary between 0.0 g Ni/t Pb from lead production to about 1,470 g Zn/t Zn from zinc production.

These Canadian trace metal emission factor data are of considerable value in enabling a comparison of the performance of Canadian smelters with estimates published in the literature. We have seen previously (Skeaff and Dubreuil, 1997) that for the relevant emission factor estimates of Nriagu and Pacyna (1988):

- three - those for Cd and Hg from lead production and for Sb from zinc production - are close to the values of ε_C
- one - Ni from Cu+Ni production - would yield an emission ~40% higher than 1993 Canadian smelter practice; and
- 21 are greater than ε_C by factors of ~2 or larger, so the range mid-points of Nriagu and Pacyna (1988), if used to estimate emissions from the Canadian primary non-ferrous metals industry, would result in overestimations by 100% and up, for the releases of:
  - ◊ As, Cu, Pb and Zn from all smelters;
  - ◊ Cd from Cu+Ni production; and
  - ◊ Hg from Zn production;
  - ◊ Mn from Cu production;
  - ◊ Ni from Pb production;
  - ◊ Sb from Pb and Cu+Ni production; and
  - ◊ Se from Cu+Ni and Zn production.

The implications of the differences between calculated and estimated emission factors in reckoning release inventories become clear in Figure 1, in which, for trace metal releases to air from Canadian smelters in 1993, we compare the total reported trace metal emissions, e_C , and those calculated, e_P , from the range mid-point emission factors of Nriagu and Pacyna (1988) and the metal production data of Table 1. (Due to scaling considerations, we have omitted from Figure 1 the estimated zinc emission, which is about 93,000 t, while the reported release is about 1,230 t.) General over-estimations of releases are readily seen. Thus it would be difficult to obtain an acceptably representative estimate of the current status of trace metal emissions for Canadian non-ferrous smelters from the emission factors of Nriagu and Pacyna (1988).

It should be noted that Nriagu and Pacyna (1988) estimate the worldwide emission factors of Zn from the non-ferrous zinc industry to lie in the range 100,000-180,000 g/t, which when multiplied by their estimated global production/consumption of 4.6 x 10^9 kg Zn/yr gives atmospheric emissions of Zn from the zinc industry in the range 460,000-828,000 t/yr, not 46,000-82,800 as per Table 2 of their paper. Using their Zn mid-range emission factor for the zinc industry then yields an annual worldwide Zn emission from the non-ferrous zinc industry of about 650,000 t, which, when added to their estimated emissions of Zn from all other sources, gives a total anthropogenic release in the range 484,250-938,700 t/yr, not 70,250-193,500 t/yr as per Table 2 of their paper. Their emission factor of Zn from the zinc industry perhaps needs revision, as it represents between 10 and 18% of all zinc produced and the associated Zn release corresponds to the refined zinc production of three commercial zinc operations.

We have previously (Skeaff and Dubreuil, 1997) compared the calculated trace metal emission factors for Canadian smelter practice with those values, ε_E , suggested by EMEP/CORINAIR (1995) to estimate non-ferrous emission inventories for four levels of emission control for lead and copper production:

lead production	**copper production**
a: improved abatement	c: impact control (≥99.9%)
b: limited abatement	d: limited control (≥99%)

Their recommended emission factors for zinc production correspond to only one level of emission control. While the EMEP/CORINAIR (1995) copper production emission factors do not include emissions for nickel production as we have calculated here, nor are their data as comprehensive as those of Nriagu and Pacyna (1988), comparisons with Canadian smelter practice are nonetheless useful. We find that the weighted average Canadian non-ferrous emission factors, ε_{wa} , for:

- lead production:
  - ◊ approach but exceed the value for "limited abatement" for Pb,
  - ◊ lie between "limited abatement" and "improved abatement" for Zn, and
  - ◊ meet or are lower than "improved abatement" for Hg and Cu; but
  - ◊ exceed "limited abatement" for As and Cd;

while there are no recommended values of ε_E for Mn, Ni, Sb and Se;

- copper (+ nickel) production:
  - ◊ approach the values for "impact control" on As and Cd;
  - ◊ lie between "limited control" and "impact control" for Cu, Ni, Pb and Zn;
  - ◊ exceed the value for Hg "impact control", although there is no "limited control" value; and
  - ◊ are significantly lower than "impact control" values for Mn, Sb and Se;

- zinc production:
  - ◊ are significantly lower, by between ~25% and almost 700%, than the suggested values for "thermal" production (presumably the roasting step of the conventional roast-leach-electrowin zinc process) for Cd, Hg, Pb, and Zn;
  - ◊ while EMEP/CORINAIR (1995) do not suggest values for As, Cu, Mn, Ni, Sb and Se.

Thus, where comparisons are possible, the emission factors for Canadian non-ferrous lead and copper+nickel production practice as calculated here are reasonably well represented by those suggested by EMEP/CORINAIR, which provides some validation for the latter. The only major exceptions are for arsenic and cadmium from lead production, and the EMEP/CORINAIR values may be for processes treating concentrates with negligibly small amounts of As and Cd, which would yield unrealistically low values of ε_E. On the other hand, the four suggested EMEP/CORINAIR values for zinc production are significantly higher than the calculated Canadian values.

Figure 2 affords a comparison of the total amounts of reported trace metal releases to air from Canadian smelters in 1993, e_C , with the totals calculated, e_E , from the two levels of emission factors suggested by EMEP/CORINAIR (1995) and the metal production data. (As with Figure 1, scaling considerations led us to omit from Figure 2 the estimate of the amount of zinc released, which is about 7,000 t, while the reported release is about 1,230 t.) If an EMEP/CORINAIR emission factor was assigned a range, we used its mid-point. Figure 2 shows that the two suggested levels of EMEP/CORINAIR emission factors will yield release estimates that lie on either side of the total reported 1993 releases of As, Cu, Ni and Pb from Canadian smelters, and yield overestimates of the reported releases of Cd, Hg, Mn, Sb and Se.

Table 2 gives the reported releases of As, Cd, Hg, Ni and Pb for 1993 as per above and for 1995 (EC, MAC-ARET, 1997), while Table 3 gives the 1993 and 1995 weighted average trace metal emission factors, ε_{wa} , of As, Cd, Hg, Ni and Pb for Canadian non-ferrous smelting calculated from these reported emission data and 1993 metal production data. Figures 3 and 4 show the total releases for 1993 and 1995. The data of Table 2 indicate that, between 1993 and 1995, some decreases in emissions occurred:

- 35% for As from zinc production
- 73% for Cd across the sector
- 60% for Hg from copper+nickel and zinc production, and
- 92% for Pb from zinc production.

The emission of nickel from copper+nickel production increased by 24%. Similar changes are reflected in the emission factors of Table 3.

CONCLUSIONS

We have used reported, publicly available, trace metal release data and production data to calculate the emission factors to air of As, Cd, Cu, Hg, Mn, Ni, Pb, Sb, Se and Zn for lead production, copper-nickel production and zinc production by the Canadian primary non-ferrous smelting industry.

With the two exceptions noted herein, our calculated weighted average emission factors, ε_C, for the Canadian non-ferrous metals industry of the mid-1990's are between approximately 40% and more than 1500 times lower than the published emission factors, ε_P, of As, Cd, Cu, Hg, Mn, Ni, Pb Sb, Se and Zn used to estimate global atmospheric release inventories for the early 1980's for the world primary non-ferrous metals industry (Nriagu and Pacyna, 1988).

Comparisons of Canadian trace metal emission factors with those suggested by EMEP/CORINAIR (1995), where possible, tend to validate the latter, except for As and Cd emission factors for lead production. Given the EMEP/CORINAIR estimated ranges of trace metal capture, perhaps 99 % of possible trace metal emissions are removed in Canadian non-ferrous practice, and this may be typical of worldwide "advanced" environmental controls of roughly the same efficiency. Similar comparisons with other non-ferrous smelting industries would be of value.

The weaknesses inherent in generalised and aggregated trace metal emission factors, and the clear necessity for accurate, currently valid, site-specific data, especially as related to metal production, become evident. Particular difficulties - such as unnecessarily burdensome environmentally-based trade barriers and regulations adding significantly to the cost of producing metals, with no measurable environmental improvement - could arise in using published but unrevised emission factor estimates in a "Sufficient Weight of Evidence" approach to substantiate our state of knowledge of trace metal emission inventories from non-ferrous smelters and other anthropogenic sources, and further to base environmental policies on. Detailed critical analysis will become possible only as more rigorous examination and efforts to achieve self-consistent data are undertaken.

The need for accurate and up-to-date measurements of atmospheric trace metal releases, and moreover, standardised analysis techniques to guide environmental protection measures, is obvious.

ACKNOWLEDGEMENTS

The authors would like to thank A.H. Webster for his critical review of, and his resulting many helpful suggestions for improvements to, the manuscript.

REFERENCES

Ashman D.W. and Jankola W.A., 1990. Recent experience with zinc pressure leaching at Cominco. In *Lead-Zinc '90, Proc. World Symp. on metallurgy and environmental control,* 119th Annual TMS Meeting, Anaheim, Calif. U.S.A., Feb. 18-21, 1990; TMS, Warrendale, Penn. (edited by T.S. Mackey and R.D. Prengaman), pp 253-275.

Biswas A.K., and Davenport W.G., 1980. *Extractive Metallurgy of Copper*, 2nd edition, Pergamon Press, Oxford, New York.

BMS (Brunswick Mining and Smelting Corporation, Bathurst, N.B., Canada). 1994a. 1994 ARET Program Action Plan: *Submission to ARET Secretariat*, Environment Canada. "ARET" in the tables; 1994b. Brunswick Mining and Smelting Corporation Annual Report 1994.

Collins M.J., McConaghy E.J., Stauffer R.F., Desroches G.J. and Krysa B.D., 1994. Starting up the Sherritt zinc pressure leach process at Hudson Bay. *J. Metals* **46** (4), 51-58.

Crowson P., 1994. Minerals Handbook 1994-1995. Bath Press, Avon, U.K.

Daum K.H. and Sneyd R.J., 1992. Gas purification for metallurgical sulphuric acid plants. *In Proc. Int. Symp. on Non-ferrous Pyrometallurgy: Trace Metals, Furnace Practices and Energy Efficiency,* 31st Annual Conference of Metallurgists of CIM, Edmonton, Aug. 23-27, 1992; Canadian Institute of Mining, Metallurgy and Petroleum, Montréal (edited by R. Bergman *et al.*), pp 85-106.

Dutrizac J.E. and Chen T.T., 1990. A chemical and mineralogical study of silver, lead and cadmium in Kidd Creek Zinc Concentrates and roaster products. In *Lead-Zinc '90, Proc. World Symp. on metallurgy and environmental control,* 119th Annual TMS Meeting, Anaheim, Calif. U.S.A., Feb. 18-21, 1990; TMS, Warrendale, Penn. (edited by T.S. Mackey and R.D. Prengaman), pp 161-191.

EC (Environment Canada, Ottawa), 1994. *Emissions from the Canadian Iron and Steel Industry in 1992*; 1995. *Summary Report of the 1993 National Pollutant Release Inventory* (NPRI). Complete data on the Internet through: HTTP:/www.ec.gc.ca/pdb/npri/index.html. 1997 http://www.npri-inrp.com/queryform.cfm or use engine to search for NPRI.

EMEP/CORINAIR, 1995. Atmospheric Emission Inventory Guidebook, second draft; *Co-operative Programme for Monitoring and Evaluation of the Long Range Transmission of Air Pollutants in Europe/The Atmospheric Emission Inventory for Europe (EMEP/CORINAIR), EMEP (European Monitoring and Evaluation Programme) Task Force on Emission Inventories*, Copenhagen. Edited by Gordon McInnes.

Eatough D.J., Eatough N.L., Hill M.W., Mangelson N.F., Ryder J., Hansen L.D., Meisenheimer R.G. and Fischer J.W., 1978. The chemical composition of smelter flue dusts. *Atmospheric Environment* **13**, 489-506.

Inco Ltd., 1994. Inco 1993 Annual Report. Toronto.

Jaques A.P., 1987. *Summary of emission of antimony, arsenic, cadmium, copper. lead, manganese, mercury, and nickel in Canada*. Environment Canada, Ottawa.

Langlois B., 1993. Les Mines Gaspé - Smelter operations. *CIM Bulletin* **86** (973), 63-67.

MAC-ARET (The Mining Association of Canada/Environment Canada) 1995, 1997. Voluntary Emissions Reduction, ARET Program. "M-A" in the tables.

MacLatchy J., 1992. . *Metals Data from Base Metal Smelters and Refineries: (As, Cd, Cr, Ni, Pb, Hg)*, Environment Canada, Ottawa. "MacL" in the tables.

McDonald A. and Verhelst D., 1994. Pyrometallurgical operations at Hudson Bay Mining and Smelting Co. Ltd. *CIM Bulletin* **87** (984), 98-102.

Mollison A.C. and Moore G.W., 1990. Zinc sulphide pressure leaching at Kidd Creek. In *Lead-Zinc '90, Proc. World Symp. on metallurgy and environmental control,* 119th Annual TMS Meeting, Anaheim, Calif. U.S.A., Feb. 18-21, 1990; TMS, Warrendale, Penn. (edited by T.S. Mackey and R.D. Prengaman), pp 277-291.

Newman C.J., Storey A.G., Macfarlane G. and Molnar K., 1992. The Kidd Creek copper smelter - an update on plant performance. *CIM Bulletin* **85** (961), 122-129.

NRCan (Natural Resources Canada, Ottawa), 1994a. Mining and Mineral Processing Operations in Canada, Report MR 239; 1994b. 1993 Canadian Minerals Yearbook: Review and Outlook

Nriagu J.O. and Pacyna J.M. 1988. Quantitative assessment of worldwide contamination of air, water and soils by trace metals *Nature* **333**, 134-139.

Nriagu J.O., 1989. A global assessment of natural sources of atmospheric trace metals. *Nature* **338**, 47-49.

St. Eloi R.J., Newman C.J. and Macfarlane G., 1994. Modifications to the process gas handling system at Kidd Creek's copper smelter. *CIM Bulletin* **87** (977), 77-85.

Skeaff, J.M., and Dubreuil, A.A., 1997. Calculated 1993 Emission Factors of Trace Metals for Canadian Non-Ferrous Smelters. *Atmospheric Environment*, **31**(10), 1449-1457.

Stubina N., Chao J., and Tan C., 1994. Recent electric furnace developments at Falconbridge (Sudbury operations). *CIM Bulletin* **87** (981), 57-61.

Thompson S.L., Morris D.L. and Firlotte R., 1987. Investigation of fluxing in a lead blast furnace. *Canadian Metallurgical Quarterly* **26** (1), 5-11.

Toop G.W., 1994. Pyrometallurgical operations of the Cominco lead-zinc smelter. *CIM Bulletin* **87** (983), 89-92.

Webster A.H., 1995. Personal communication. Ottawa.

TABLE 1: Calculated trace metal emission factors for Canadian non-ferrous smelters, 1993

Canadian non-ferrous smelter and tonnes of metal produced			arsenic		cadmium		copper		mercury		manganese		sources of production capacity data
			e	ε_C	e	ε_C	e	ε_C	e	ε_C	e	ε_C	Thompson and Morris (1987);
BMS	74000	Pb	5.50 NPRI	74.3	3.20 NPRI	43.2	0.34 ABR	4.59	0.06 M-A	0.81			BMS 1994 Annual Report (BMS, 1994b)
Cominco			7.00 NPRI		3.00 NPRI		1.00 NPRI		1.60 NPRI		0 NPRI		
	100000	Pb	2.06	20.6	0.88	8.8	0.29	2.94	0.47	4.71	0.0	0.00	Toop (1994)
	240000	Zn	4.94	20.6	2.12	8.8	0.71	2.94	1.13	4.71	0.0	0.00	Toop (1994)
Falconbridge (Kidd)			0.386 NPRI		0 NPRI		53.6 NPRI		0.02 M-A		0 NPRI		
	120000	Cu	0.19	1.54	0.00	0.0	52.75	440	0.01	0.08	0.0	0.00	St. Eloi et al. (1994).
	130000	Zn	0.20	1.54	0.00	0.0	0.85	6.52	0.01	0.08	0.0	0.00	Newman et al. (1992)
Falconbridge (Sudbury)													
	60000	Cu+Ni	0.2 M-A	3.33	2.5 M-A	41.7	7.14 NPRI	119			0.129 NPRI	2.15	Stubina et al. (1994)
HBMS			27 M-A		67 M-A		170.4 NPRI		8 M-A				
	76000	Cu	14.798	195	36.7	483	169.1	2225	4.38	58			McDonald and Verhelst (1994)
	62667	Zn	12.202	195	30.3	483	1.33	21.2	3.62	58			Collins et al. (1994)
Noranda													
CEZ	230000	Zn	0 NPRI	0	0.349 NPRI	1.5	0.790 NPRI	3.43	0.28 M-A	1.24	0.170 NPRI	0.74	NRCan (1994a, 1994b)
Gaspe	66000	Cu	18.96 NPRI	287	0 NPRI	0.0	1.7 NPRI	25.8	0 NPRI		0 NPRI	0.00	Langlois (1993)
Horne	215000	Cu	23.0 NPRI	107	5.4 NPRI	25.1	149 NPRI	693	0.27 M-A		0.014 NPRI	0.07	NRCan (1994a, 1994b)
totals	281000	Cu	41.96 NPRI	149	5.4	19.2	150.7	536	0.27		0.014	0.05	
	230000	Zn	0	0	0.349	1.5	0.79	3.43	0.28	1.24	0.17	0.74	
Inco													
Copper Cliff *	193928	Cu+Ni	8.51 NPRI	43.9			211.8 NPRI	1092					Inco Annual Report for 1993 (1994)
Thompson	56527	Cu+Ni	6.16 NPRI	109			22.2 NPRI	392					Inco Annual Report for 1993 (1994)
totals	250455	Cu+Ni	14.7	58.6			234.0	934					
TOTALS				ε_{wa}		ε_{wa}		ε_{wa}		ε_{wa}		ε_{wa}	
	174000	Pb	7.56	43.4	4.08	23.5	0.63	3.64	0.53	3.05	0.0	0.0	
	787455	Cu+Ni	71.8	91.2	44.6	56.7	614	779	4.39	5.58	0.14	0.18	
	662667	Zn	17.3	26.2	32.7	49.4	3.67	5.54	5.0	7.60	0.17	0.26	
			96.7		81.4		618		9.96		0.31		

			nickel		lead		antimony		selenium		zinc		
			e	ε_C	e	ε_C	e	ε_C	e	ε_C	e	ε_C	Thompson and Morris (1987);
BMS	74000	Pb	0 Mod.	0.0	21.7 NPRI	293					3.2 ABR	43.2	BMS 1994 Annual Report (BMS, 1994b)
Cominco			0 Mod.		83.0 NPRI		12.0 NPRI				35.0 NPRI		
	100000	Pb	0.00	0.0	69.9	699	3.53	35.3			4.42	44.2	Toop (1994)
	240000	Zn	0.00	0.0	13.1	54.4	8.47	35.3			30.6	127	Toop (1994)
Falconbridge (Kidd)			0 Mod.		2.94 NPRI				0.164 NPRI		86.18 NPRI		
	120000	Cu	0.00	0.0	1.41	11.8			0.079	0.66	11.1	92.4	St. Eloi et al. (1994).
	130000	Zn	0.00	0.0	1.53	11.8			0.085	0.66	75.1	578	Newman et al. (1992)
Falconbridge (Sudbury)													
	60000	Cu+Ni	10.33 NPRI	172.2	17.63 NPRI	294					7.304 NPRI	122	Stubina et al. (1994)
HBMS			2 Mod.		522 M-A						864 NPRI		
	76000	Cu	1.10	14.4	286	3764					141	1854	McDonald and Verhelst (1994)
	62667	Zn	0.90	14.4	236	3764					723	11539	Collins et al. (1994)
Noranda													
CEZ	230000	Zn	0 Mod.	0.0	1.5 NPRI	6.52			0.015 NPRI	0.07	143 NPRI	621	NRCan (1994a, 1994b)
Gaspe	66000	Cu	0.85 NPRI	12.9	20.04 NPRI	304	0 NPRI	0.0	2.1 NPRI	31.8	3.37 NPRI	51.1	Langlois (1993)
Horne	215000	Cu	0.725 NPRI	3.4	215 NPRI	1000	1.15 NPRI	5.3			86 NPRI	400	NRCan (1994a, 1994b)
totals	281000	Cu	1.575	5.6	235	836	1.15 NPRI	4.1	2.599 NPRI	9.25	89	318	
	230000	Zn	0	0.0	1.5	6.52			0.015	0.07	143	621	
Inco									applies to copper refinery				
Copper Cliff *	193928	Cu+Ni	334 NPRI	1722	111.37 NPRI	574			0.239 NPRI	1.2			Inco Annual Report for 1993 (1994)
Thompson	56527	Cu+Ni	150 NPRI	2650									Inco Annual Report for 1993 (1994)
totals	250455	Cu+Ni	484	1932	111.37	445			0.239	1.0			
TOTALS				ε_{wa}		ε_{wa}		ε_{wa}		ε_{wa}		ε_{wa}	
	174000	Pb	0.00	0.0	91.6	527	3.53	20.3			7.62	43.8	
	787455	Cu+Ni	497	631	652	827	1.15	1.5	2.92	3.7	249	316	
	662667	Zn	0.90	1.36	252	380	8.47	12.8	0.10	0.2	972	1466	
			498		995		13.2		3.02		1228		

e = reported Canadian emission, t (stack plus fugitive)
ε_C = calculated Canadian emission factor, g/t
ε_{wa} = weighted average trace metal emission factor for the Canadian non-ferrous smelting industry, g/t
* comprised of a smelter, a copper electrowinning and refining plant, and a nickel carbonyl refinery

TABLE 2: Reported 1993 and 1995 trace metal emissions, e in t, of As, Cd, Hg, Ni and Pb for Canadian non-ferrous smelters.

		arsenic		cadmium		mercury		nickel		lead	
		1993	1995	1993	1995	1993	1995	1993	1995	1993	1995
metal	Pb	7.56	6.39	4.08	3.15	0.53	0.53	0	0.0	91.6	82.1
produced	Cu+Ni	71.8	71.2	44.6	13.9	4.39	2.49	497	618	652	507
	Zn	17.3	11.2	32.7	4.7	5.00	1.28	0.9	0.07	252	19.3
	TOTALS	96.7	88.8	81.4	21.7	9.92	4.30	498	618	996	608

TABLE 3: Calculated 1993 and 1995 weighted average trace metal emission factors, ε_{wa} in g/t, of As, Cd, Hg, Ni and Pb for Canadian non-ferrous smelters

		arsenic		cadmium		mercury		nickel		lead	
		1993	1995	1993	1995	1993	1995	1993	1995	1993	1995
metal	Pb	43.4	36.7	23.5	18.1	3.1	3.0	0	0.0	527	472
produced	Cu+Ni	91.2	90.5	56.7	17.6	5.6	3.2	631	785	827	644
	Zn	26.2	18.6	49.4	7.8	7.6	2.1	1.4	0.12	380	32.1

Figure 1:

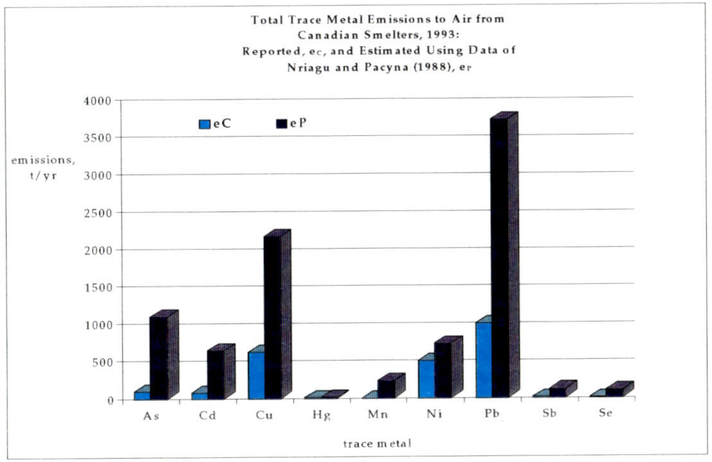

Figure 2:

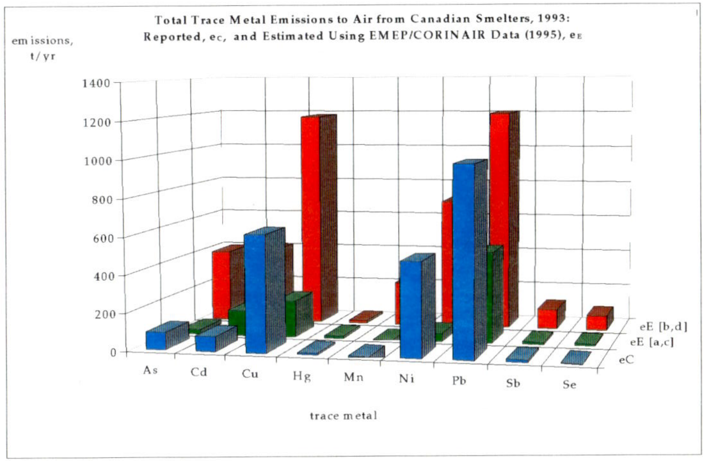

Figure 3:

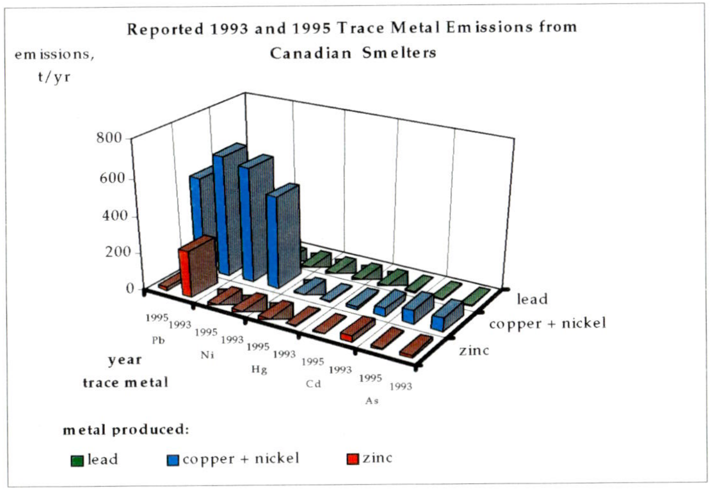

Figure 4:

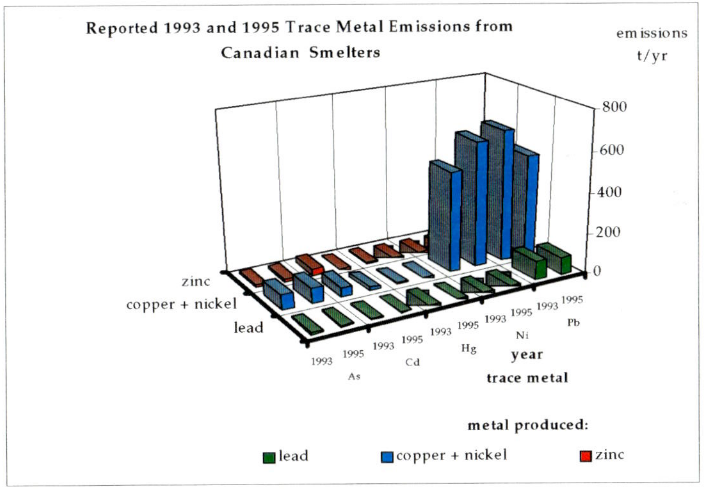

Metal commodities and products

Marketing challenges confronting metals and products

L.J. Surges
Noranda Inc.
1 Adelaide St. E., suite 2700
Toronto, Canada M5C 2Z6

ABSTRACT

Chemicals management policies do not generally recognize the special character and properties of metals, alloys and inorganic metal compounds. Expert judgement is needed to avoid misleading classification and labeling and excessive risk reduction efforts. Attention is shifting from production processes to products, and metal producers are increasingly expected to assume some responsibility for health and environmental risks. Producers must become more proactive in assessing effects of metal production and products and communicating benefits and risks to policy-makers and the public. Producers must particularly market the advantages of metals over competing materials to product designers and consumers.

RÉSUMÉ

Les politiques de gestion des produits chimiques ne reconnaissent habituellement pas la nature et les propriétés spéciales des métaux, des alliages et des composés métalliques inorganiques. Une expertise est requise afin d'éviter les classifications et les étiquetages trompeurs, ainsi que les efforts excessifs de réduction des risques. L'attention passe des procédés de production aux produits eux-mêmes et l'on s'attend de plus en plus à ce que les producteurs de métaux assument une certaine part de responsabilité quant aux risques pour la santé et l'environnement. Les producteurs se doivent d'être plus proactifs dans l'évaluation des effets des métaux et de leur production, et dans la transmission des avantages et des risques aux décideurs et au public en général. Ils doivent particulièrement vanter les avantages des métaux (par rapport aux matériaux concurrents) aux concepteurs de produits et aux consommateurs.

Proceedings of
Metals and the Environment: An International Symposium
Edited by Jim Skeaff
The Metallurgical Society of CIM
100th CIM Annual General Meeting, 1998

INTRODUCTION

The unique properties of metals, alloys and advanced metallic materials allow product designers to create and meet consumer demand for products that offer new or improved service at lower cost. The social, economic and other benefits afforded by metals have been obscured by urbanization, a growing service sector and new knowledge-based industries, while concerns about health and the environment have intensified. Governments, businesses and consumers respond to these concerns. Initiatives for the sound management of chemicals at the national, regional and international level do not fully account for differences between inorganic metal compounds and organic chemicals. Metals are increasingly viewed as an international, regional or global problem to be resolved through cooperative action. Attention is shifting from production processes to products, and metal producers are increasingly expected to assume some responsibility for health and environmental risks related to products. Factors driving environmental policy development, reactions by producers and the need for a strategic response are examined.

USE OF SCIENCE

Minerals, metals and other materials are the milestones by which historians describe early human progress and remain of fundamental importance, despite social development and a trend toward reduced material intensity. New products offering new possibilities can also pose new risks. Chemical fertilizers and pesticides improved life expectancy and other social measures through eradication of disease and increased food production, while Polychlorinated biphenyls (PCBs) facilitated distribution and use of electricity. Risks to health and the environment received less consideration. Products were presumed to be safe unless proven otherwise. Industry research and development focused on product and process development, with most environmental monitoring being performed by government scientists or academics.

Early environmental concerns were characterized by actual impacts and a high degree of certainty regarding the cause. Sources were typically manufacturing facilities that could be readily identified, and legislatures and governments responded by enacting technology-based pollution control legislation. Examples include the *Metal Mining Liquid Effluent Regulations* under the *Fisheries Act* and the *Asbestos Mines and Mills Release Regulations* and the *Secondary Lead Smelter Release Regulations* promulgated under the *Clean Air Act* and continued under the *Canadian Environmental Protection Act*.

Causes of other environmental impacts were less apparent. For example, eutrophication of rivers and lakes resulted from phosphate and other nutrient releases in wastewater and acidification resulted from acid precipitation caused by acid gas emissions. Consumer use and disposal as well as industrial sources were recognized as sources of pollution. Reduced hatching success in fish-eating birds of prey was associated with thinner egg shells which were in turn linked to biomagnification of Dichloro-diphenyl-trichloretan (DDT), a pesticide used to control mosquito populations and reduce the incidence of malaria. Similarly, traces of PCBs were found in marine mammals in the Arctic and in human breast milk. The effects of chronic exposure were unknown, but scientists warned that exposures would increase and the potential effects could be grave. These impacts were profoundly disturbing: if wildlife in remote, pristine regions were not safe from the harmful effects of toxic chemicals, how could the general population and particularly the most vulnerable members of society be protected from the harmful effects of chemicals? Domestic product measures were implemented, but concerted international action was also necessary and new policy frameworks were needed to protect society from similar threats in the future. Concern about toxic chemicals was characterized by the potential for serious adverse impacts and a high degree of complexity and scientific uncertainty. The guiding philosophy was that a product or substance would be presumed to be unsafe until proven otherwise, and the burden of proof shifted to industry.

This preventative approach is reflected in the *Canadian Environmental Protection Act*. Lead and mercury were added to the List of Toxic Substances and the Act included which included a mechanism for selection of existing substances to be added to the Priority Substances List for assessment based on available data and information. Enabling provisions authorized a process for new substances notification, testing and review to be established by regulation, and the *New Substances Notification Regulations* require industry to provide extensive data and information concerning new substances.

MANAGEMENT INITIATIVES

Several initiatives have been taken for the sound management of chemicals. Examples include Canada's *Toxic Substances Management Policy* (TSMP), a draft Canadian Council of Ministers of the Environment (CCME) *CCME Policy for the Management of Toxic Substances*, a regional initiative under the North American Agreement on Environmental Cooperation (NAAEC) and international initiatives within the European Union (EU) and the Organization for Economic Cooperation and Development (OECD). Most seek to manage risks associated with chemicals throughout their life cycle and utilize screening criteria to select substances for risk assessment and possible risk management. Scientists examined the behaviour of environmental contaminants and observed that all were persistent, bioaccumulative and toxic.

Most chemicals management policies do not recognize the special character and properties of metals, alloys and inorganic metal compounds. Metals have been used for thousands of years, but gaps exist in our understanding of the environmental behaviour, transport, fate and effects of smelter emissions and end-of-life products. In the absence of such knowledge, decisions may be based on conservative science policies and lead to overestimation of risks and excessive regulatory controls and costs.

Domestic Policies

Criteria are incorporated in the federal TSMP and the implementation notes for the draft CCME policy. *The Metals and Minerals Policy of the Government of Canada* builds on the TSMP and recognizes that naturally-occurring inorganic substances, such as minerals and metals, behave differently than synthetic organic chemicals in the environment and, as a consequence, require different risk management approaches. The CCME is consulting stakeholders on a draft policy to support coordination of actions by federal, provincial and territorial governments. The draft policy recognizes that ". . . many toxic substances can be properly managed using the pollution prevention approach. A select few that are persistent, bioaccumulative and toxic pose an unreasonable and otherwise unmanageable risk and the pollution prevention goal must be virtual elimination.". In both cases, decision-makers consider the origin of releases in determining the appropriate management goal. Both policies apply the same criteria to inorganic metal compounds and organic pollutants but explicitly provide that elements and naturally occurring inorganic compounds are not candidates for virtual elimination. Social, economic and technological factors are considered in determining the schedule for virtual elimination of Track 1 substances. For Track 2 substances, the potential for releases is examined for each stage of the product life cycle. Product management measures such as bans or phase-outs are a last resort for substances that pose unreasonable and otherwise unmanageable risks to human health or the environment.

Harmonization of Classification and Labeling

Harmonization of classification and labeling of chemicals by the year 2000 was identified by the 1992 United Nations Conference for Environment and Development as an area requiring international cooperation. The OECD accepted a leadership role in examining acute and chronic health hazards and environmental hazards, using EU criteria for the classification "dangerous to the environment" as a basis for discussion. Criteria of persistence, toxicity and bioaccumulation are applied to organic and inorganic compounds alike, resulting in misleading classification and labeling of some minerals, metals, alloys and inorganic metal compounds. Work to date recognizes that alloys should not be treated as "mixtures" of the constituent elements and may lead to reconsideration of the EU approach to the classification and labeling of metals, alloys and inorganic compounds.

Metals are elements. Native metals occur in nature, but most metals occur as other minerals that are thermodynamically stable under geological or environmental conditions. Criteria for persistence are relevant for synthetic chemicals and indicate that degradation of the substance will be slow, extending the duration of exposure and increasing the potential for bioaccumulation. Persistence is not a useful criterion for metals, alloys and inorganic metal compounds that are intrinsically not degradable.

Toxicity is of interest for all substances. Whether or not a substance is toxic to biota depends on the chemical and biological species present and the concentration of the chemical species. Evaluations should consider the physical form and chemical speciation of a mineral, metal, alloy or inorganic metal compound since the absence or occurrence of toxic effects depends upon those factors and on the test organism, dose and duration of exposure. All life forms developed in the presence of metals, and have developed mechanisms to regulate uptake at the cellular level. Cobalt, copper, iron, manganese, molybdenum and zinc as well as the metalloid selenium are essential for plant or animal health. Other metals are not essential and organisms do not suffer from deficiencies, but are able to regulate metal uptake. Chronic exposure does not necessarily result in adverse effects on health. Toxicity data is not available for most chemical and biological species: most laboratory test data is based on the administration of a soluble metal salt. For sparingly soluble inorganic metal compounds, transformation to a soluble and bioavailable form may control the rate of uptake by the organism. Exposure to a particular dose in the form of a less soluble compound results in reduced uptake and a lower probability of adverse effects. Risk assessments that do not take account of solubility and bioavailability overestimate risk. Conservative assumptions and margins of safety may lead to risk management actions that produce little or no benefit, or may result in adverse effects if they cause deficiency.

Bioaccumulation is broadly defined as the net retention of a substance by an organism over time and is a process rather than an effect. It is a necessary process for essential metals, provided that uptake is not excessive. Bioaccumulation is not an intrinsic property of a substance but depends on exposure, speciation, solubility, bioavailability and organism response. It should be considered at the risk assessment stage, but is not a useful screening criterion unless data is available that is relevant to the particular substance, organism and environment.

NAAEC Sound Management of Chemicals Initiative

Council Resolution #95-5 was adopted by the Council of Ministers of the Commission for Environmental Cooperation (CEC), established under the NAAEC, at the Council's second regular meeting in October 1995. The resolution recognizes that cooperative actions for the sound management of chemicals are needed to protect and improve the environment and to achieve sustainable development and commits the governments of Canada, the United States and Mexico to cooperate to improve the sound management of chemicals in North America. The resolution gives priority to the management and control of substances of mutual concern that are persistent, bioaccumulative and toxic, but also allows for cooperation on a broader scale for the sound management of chemicals. The resolution required North American Regional Action Plans (NARAPs) to be developed for PCBs and three additional substances. The three additional substances selected for early cooperative action were DDT, chlordane and mercury.

The resolution required a process and criteria to be developed to help identify potential candidate substances for future NARAPs. Toxicity data is included in the nomination dossier, but is not an evaluation criterion. The screening evaluation criteria include the existence of an available and acceptable risk assessment and a judgement on measured/predictive data sets that differ for persistent organic pollutants (POPs) and metals, minerals and naturally occurring substances. Bioaccumulation and persistence are considered for POPs, while bioaccumulation and bioavailability are considered for metals, minerals and naturally occurring substances. Quantitative criteria are recommended for POPs, including organometallic compounds such as alkyl lead and methyl mercury. Expert judgement takes precedence in interpreting the four criteria for all substances and is particularly important for metals, minerals and naturally occurring substances, for which quantitative criteria are not recommended. The Task Force report includes a synopsis of recent international meetings on metal classification.

INDUSTRY POSITIONING

Chemical producers have positioned themselves as a proactive industry that accepts responsibility for managing risks associated with its facilities and products through Responsible Care® and other initiatives. Metal producers also emphasize the need for sound science and risk-based decision-making but place greater importance on competitiveness.

Both industries have achieved substantial release reductions. The mining industry has reacted to specific threats, but has not adequately funded research into the environmental effects of its activities and products. The industry has been unable to formulate a common approach to product stewardship and producers have not consistently supported user industries under pressure to substitute competitive materials or reduce the metal content of products.

PUBLIC PERCEPTIONS

Industry operates with the consent of the public. Results of a survey (Corporate Research Associates Inc., 1997) indicate that Canadian attitudes towards the mining industry were somewhat more positive than in 1993. With the exception of the discussion of safety and environmental risks, the following observations are based on that survey.

Economic Importance

Canadians continue to view resource industries, primarily forestry, as the most important sector of the Canadian economy except in Ontario, where manufacturing is increasingly viewed as the country's top sector. The importance placed on the service sector is largely defined by community size, with those living in urban areas most likely to identify this sector as number one.

Canada's mining industry has increased its public profile since 1993, when the previous survey revealed a lack of awareness and an underestimation of the industry's economic contribution. The industry's rising profile is reflected by the fact that at least seven in ten Canadians now agree that it is an important source of jobs and contributes to the country's international image abroad. Canadians generally expect resource industries to grow rather than shrink in importance over the next 10 years. The mining industry has recorded the most significant improvement in this area, although it lags most other sectors.

Canadians are divided concerning the future role of mining in the Canadian economy. While 52% anticipate a stronger role, up 9% for the strongest increase of any sector, one-third believe that the sector will be less important 10 years from now. Anticipation of improving prospects have increased most substantially in the Prairies (63%, up 15) and British Columbia (52%, up 19) as well as among Canadians in the top income bracket (52%, up 21). Residents of Québec continue to be least optimistic and are evenly divided with 41% expecting the industry to grow more important and 39% who think it will become less so. Although there is no clear consensus about the future growth of the mining industry, there is little evidence that it is seen as a "sunset" industry destined to fade away in the shadow of an emerging knowledge-based economy.

Environmental Impacts

The public continues to see resource industries as significantly damaging the environment. Mining activity is not as widely viewed as causing significant environmental damage relative to most other resource industries, but continues to be seen as disruptive to nearby land uses. Opinions about the acceptability of industry damage to the environment remain divided, although concerns have moderated somewhat. The public is a bit more likely than before to believe that the sector is demonstrating a commitment to reduce the environmental damage resulting from its operations, but the small magnitude of the trend indicates that industry has made little progress in convincing the public of its commitment to sound environmental practices. Although Canadians remain divided about the extent and acceptability of environmental damage, they increasingly believe that reducing such damage makes sound business and economic sense.

Science and Technology

Canadians are as likely to see science and technology as having a critically important role in the sustainable development of resources as in contributing to economic prosperity and improving the overall quality of life, but mining continues to have a "low-tech" public image. The increase in the image of mining as "high-tech" has taken place mostly in Ontario and the Prairies. Across the country as a whole it is younger Canadians and those with the most education who have shown the most movement in how the industry is viewed.

Safety and Environmental Risks

Perceptions that mining is a hazardous occupation are reinforced by the Westray disaster and the ensuing criminal trials, civil inquiry and media interest. Environmental incidents also capture the imagination. Mine safety and environmental incidents will be raised whenever a new mine development is proposed. Mine operators need to assess and communicate impacts and risks to workers and communities and develop and implement appropriate frameworks to manage those risks. The industry also needs to document and communicate examples of best practices.

NEW PARADIGMS

To maintain and enhance public understanding of the benefits of metal products and social acceptance of mining, in Canada and abroad, the industry must challenge established paradigms related to metals and competing materials.

Non-Renewable or Abundant Resource?

The distinction between renewable and non-renewable resources reflects concern expressed by the Club of Rome that society was rapidly running out of resources. Distinctions should be made between resources that are scarce or abundant, and recyclable or not recyclable. Minerals are relatively abundant components of the earth's crust. Resources can be classified as known and unknown and economic and uneconomic. Ore reserves are resources that are known and economic, reflecting metal prices and technology as well as exploration activity. A particular ore body will eventually be depleted, but new metal entering service can be recycled repeatedly. Domestic ore reserves for some metals are declining as exploration shifts offshore, but world reserves are more than keeping pace with demand while exploration technology improves and production costs continue to decline. Declining long term metal prices do not indicate a market perception that supplies of metals are likely to become scarce.

Many competing materials are derived from petroleum products that are comparatively scarce. Recycling rates are limited by technical and economic factors and product life cycles for many applications are relatively short. As a result, energy resources must be depleted and competing materials are lost to landfills and incinerators.

Recyclable or Recycled?

Metals and alloys can be recycled repeatedly without loss of performance, although care must be taken to ensure that alloy specifications are met. For example, secondary aluminum is widely used in cast applications while wrought alloys rely more on primary aluminum. Recycling rates vary by material and application. For the lead-acid battery, studies for the Battery Council International indicate that recycling rates increased from 88.6% for 1987 to 98.2% for 1994 despite the low value of the commodity. Studies of municipal solid waste streams do not account for apparent losses of batteries from the system, possibly due to temporary storage. Recycling rates for other metals are more difficult to determine due to the diverse number of product applications and extended product life cycles. Production data indicate the importance of recycling to some degree, but the percentage of secondary production of a metal is generally lower than the recycling rate due to market growth. Scrap metals are commodities and can be readily marketed, although new markets may need to be developed when new technology is implemented.

Other material competitors recognize that the public expects product manufacturers and material producers to play a role in diverting industrial, commercial and post-consumer waste from final disposal to recycling. Plastic and other material types are now identified by pictograms, but recycling is limited by technical and economic issues related to sortation and marketing and relies largely on source separation. Social benefits in terms of disposal cost avoidance can be demonstrated, but depend on consumer cooperation and are reduced by costs of collection, transfer, storage and marketing. Recycling rates vary widely according to local programs and participation.

Metals are widely recycled, while few competing materials are recycled to a comparable degree. Metals recovery from end-of-life vehicles is highly advanced in North

America, and regulatory and non-regulatory initiatives in Europe and North America are expected to significantly increase recovery rates from end-of-life vehicles, electronic products, white goods and municipal solid waste incinerator grate siftings.

Ecological labeling is a means to communicate information to consumers, but the process must be scientifically sound and transparent. Labeling on the basis of material origin is a potential threat. Canada is a leading producer and exporter of refined metals. It would not be environmentally preferable or economically feasible to import sufficient quantities of recyclable materials to market metals on the basis of a minimum recycled content. Moreover, there is no valid environmental need to stimulate demand for products manufactured from recycled metals. Efforts should focus on life cycle management to reduce material losses and ensure that losses do not pose unacceptable risks.

Biodegradable or Persistent?

As unanticipated effects of pesticides and other synthetic organic chemicals came to light, scientists considered that biodegradability would be a desirable attribute for synthetic chemicals that are dispersed in the environment. Over time, biodegradability has become associated with "not persistent". Manufacturers of plastic packaging reformulated products to encourage degradation when exposed to ultraviolet light. This modification reduced visible evidence of plastics in the environment, while packaging reductions and recycling programs reduced material flows to landfills and incinerators. The criterion of persistence also focused attention on dispersion of metals in the environment, since metals are inherently persistent and non-biodegradable.

Persistence is a useful hazard criterion for synthetic organic chemicals that are dispersed in the environment but is not an appropriate criterion for metals, alloys and inorganic metal compounds. Metal producers and users must convince policy-makers, product designers and consumers that persistence is a necessary and desirable characteristic for materials used in durable, reliable products that can be repeatedly recycled, reducing waste and conserving natural resources.

MARKETING CHALLENGES

Corporate Commitment and Priority

The Canadian public remains skeptical of industry's commitment to sound environmental management. Support for regulatory reform and non-regulatory approaches will remain weak unless the industry communicates its commitment and progress to the public more effectively. Ongoing efforts will be required to continually improve environmental performance by managing significant environmental aspects associated with the process life cycle, including exploration, extraction, processing, smelting, refining, waste management, decommissioning and site rehabilitation. Integrated producers must understand product life cycle management techniques, including life cycle analysis, design for the environment and the application of risk assessment and risk management approaches, in order to support users and identify and proactively address any application that poses excessive risk. The industry must assume leadership in addressing health and environmental issues related to minerals and metals. Credibility of industry release reduction and other environmental claims is also an issue.

Consumers

Manufacturers and governments respond to public demands, whether or not consumer actions are well-informed. For example, public opposition to clamshell styrene wrappings for McDonald's hamburgers did not accurately reflect the environmental impacts associated with the production and disposal of alternative types of packaging. Similarly, if consumers demand metal-free products manufacturers of consumer products will respond. Consumers must be informed that metals can be and are extensively recycled and play a vital role in recycling of post-consumer wastes for which source separation is essential, including end-of-life batteries. Consumers need to be convinced that metal content and origin are not indicative of risks or environmental burdens and that incidental losses to landfills or incinerators do not pose a threat to health or the environment.

Partnerships

Customers and product manufacturers are the link between metal producers and consumers. Producers need to develop new partnerships with metal users, or will learn too late that a decision has been made to substitute a competing material. Increasingly, producers will be expected to collect and share data concerning energy consumption, releases and other environmental aspects of their businesses and to promote end-of-life recycling advantages of metals over competing materials.

Product manufacturers will also become a more important link in the recycling chain in the future. With its expertise in the design and operation of physical separation processes, the mining industry is in a position to develop and implement new recycling infrastructure in partnership with manufacturers of consumer products and public or private sector waste management operators. Some of the mines of the future are located in our backyards, and will be strong competitors producing a variety of recyclable materials that can be converted to refined metals at low cost.

Metal producers need to work with the Government of Canada to develop and implement strategies that promote the Safe Use Principle. Such strategies could include, for example, adoption of stewardship programs based on the life-cycle approach in cooperation with user industries and other partners.

Recycling

Consumers may have a preference for products manufactured from recycled materials, but recycling has an image problem. It's not difficult to understand why: many people can recall seeing unsightly "junk yards", some of which have required costly remediation. Few would welcome a recycling facility in their back yard, although fewer would want either a landfill or an incinerator.

Recycling also suffers for two other reasons. Some recyclable materials cannot be readily marketed, particularly when prices are weak. Additionally, there have been cases where irresponsible profiteers have imported or exported wastes for final disposal under the guise of recyclable materials, either for lower cost disposal in an environmentally sound facility or for disposal in an uncontrolled manner in an underdeveloped country. As a result, complex domestic and international movement controls based on hazard criteria label a wide range of metal-bearing recyclable materials as "hazardous waste".

The Minerals and Metals Policy of the Government of Canada commits the government to follow a risk-based approach to the regulatory management of recyclable metals and metal-bearing materials and to promote common approaches to the definition of waste to differentiate between metal-bearing recyclable materials destined for recovery operations and wastes destined for final disposal. The Policy anticipates three types of movement controls:

- normal commercial controls for recyclable metals and metal-bearing materials that do not exhibit hazard characteristics;
- appropriate regulatory controls for recyclable metals and metal-bearing materials that exhibit hazard characteristics, but do not need to be managed as if they are hazardous wastes, and whose risks to human health and the environment can be soundly managed under conditions of normal use; and
- hazardous waste-like regulatory controls for recyclable metals and metal-bearing materials that exhibit hazard characteristics and that, despite management, continue to pose unreasonable risks to the environment or where there is a history of mismanagement.

Metal producers must work in partnership with governments and other stakeholders to develop and implement a risk-based approach to the regulatory management of recyclable metals and metal-bearing materials.

REFERENCES

CORPORATE RESEARCH ASSOCIATES INC., 1997. Tracking Survey of Canadian Attitudes Towards Natural Resources Issues, 1997. Report prepared for Natural Resources Canada

Metals and their behaviour in the environment

Iain Thornton PhD, DSc, FGS
Director, Environmental Geochemistry Research Group
Centre for Environmental Technology
Imperial College of Science, Technology and Medicine
London, United Kingdom

ABSTRACT

The topic covered by this paper is extremely broad and it will not be possible to present a comprehensive review of this complex subject. It is, however, aimed to draw attention to those aspects of metals in the environment that are important to underpin the risk assessment process on which policy and regulatory issues are determined. The sections below are based on two previous publications on this subject - the monograph - Metals in the Global Environment: Facts and Misconceptions (Thornton, 1995) and Risk Assessment Related to Metals: The Role of the Geochemist (Thornton, 1997).

Proceedings of
Metals and the Environment: An International Symposium
Edited by Jim Skeaff
The Metallurgical Society of CIM
100th CIM Annual General Meeting, 1998

SOURCES OF METALS IN THE ENVIRONMENT

Metals are natural components of rocks, soils, air, water and biological materials, and are present in varying amounts in all environmental media. The prime natural source of metals in the environment is the weathering of rocks. These range widely in their metal content. For example, course-grained sandstones contain very small amounts of metals such as copper and cobalt. On the other hand, marine black shales frequently contain large amounts of molybdenum, selenium and cadmium, while ultrabasic rocks such as serpentine may contain significant quantities of chromium and nickel.

Soils tend to reflect the chemical compositions of their geologic parent materials. Thus, soils developed from coarse-grained sediments, such as sandstones, and from acid igneous rocks, such as rhyolites and granites, tend to contain smaller amounts of the nutritionally essential metals copper, cobalt and zinc, than those derived from fine-grain sedimentary rocks, such as clays and shales, and from basic igneous rocks. Potentially toxic amounts of metals in soils may be derived from the weathering of metal-rich source rocks. Typical ranges of metal concentrations in soils in the British Isles are listed in Table 1.

The element molybdenum serves as a useful example. The mean content of Mo in British surface soils has been reported as 1 μg/g (Swaine and Mitchell, 1960), and the majority of soils contain less than 5 μg/g Mo. However, much higher concentrations are frequently found in soils developed from marine black shales. Analysis of several thousand soils at Imperial College from England, Wales and Ireland has shown that those derived from black shales of Cambrian, Ordovician, Silurian, Jurassic, Carboniferous, Cretaceous, and Recent age contain from 1 to 100 μg/g Mo, while the majority of soils derived from other shales, clays, limestones, sands, sandstones and granites contain less than 2 μg/g Mo (Thornton, 1977). Naturally occurring elevated concentrations of molybdenum in soils influence the composition of pasture herbage which in turn results in excess molybdenum ingestion by grazing animals causing a deficiency of copper. In Britain alone some 810,000 hectares of agricultural land are affected (Thomson et al, 1972). Similar types of soil parent material occur in many parts of the world and the impact has as yet to be assessed. Although industrial sources of molybdenum, particularly arising from the emissions from metallurgical plants, may result in contamination of soils, the areas affected are far smaller than those relating to "natural contamination".

Pedogenetic (soil-forming) processes modify the influence of parent materials on the soils' total metal content and the mineral and chemical forms in which the metals are found. Once again, the processes involved are complex and interactive. Gleying, leaching, surface organic matter accumulation and podzolization act together with soil properties such as reaction (pH) and redox potential (Eh) to mobilize and redistribute metals both within individual soil profiles and between neighbouring soils.

Natural sources of metals in the atmosphere arise from the eruption of volcanoes and circulation of wind-blown dust and of ocean-derived water particles. Unpolluted freshwaters, used for both drinking and irrigation, usually contain very small quantities of metals. However, in some specific localities, waters may contain elevated concentrations of metals; for example, arsenic may be leached into groundwaters out of subsurface geological materials such as pyrite.

A significant proportion of metals in the surface environment arises as a result of man's activities. These include emissions from metal mining, smelting and refining, power generation, manufacturing and transport, and accumulation arising from urban and household activities,

TABLE 1: Metals in Soils derived from Normal and Geochemically Anomalous Parent Materials in Britain

	Typical normal range in soil (ppm)	Metal-rich soils (ppm)	Sources	Possible effects
As	< 5-40	up to 2500	Mineralization	Toxicity in plants and livestock, excess in food crops
		up to 250	Metamorphosed rocks around Dartmoor	
Cd	< 1-2	up to 30	Mineralization	Excess in food crops
		up to 20	Carboniferous black shale	
Cr	15-500	up to 3500	Ultrabasic rocks in Scotland	No known effects
Cu	2-60	up to 2000	Mineralization	Toxicity in cereal crops
Hg	0.008-0.19	1-7	Mineralization	No known effect
Mo	< 1-5	10-100	Marine black shales of varying age	Molybdenosis or molybdenum-induced hypo-cuprosis in cattle
Ni	2-100	up to 8000	Ultrabasic rocks in Scotland	Toxicity in cereal and other crops
Pb	10-150	1% or more	Mineralization	Toxicity in live-stock; excess in foodstuffs
Se	< 1-2	up to 7	Marine black shale in England & Wales	No effect
		up to 500	Namurian shale in Ireland	Chronic selenosis in horses and cattle
Zn	25-200	1% or more	Mineralization	Toxicity in cereal crops

waste disposal to landfill and by incineration, and the use of agricultural materials, including pesticides, fertilizers, and the utilization/disposal of sewage sludge on farmland. Industrial releases of metals are mostly confined to regions immediately adjacent to the facilities in question. However, emissions from fossil fuel combustion, transport and the use of agricultural supplements are diffuse. The majority of metal emissions accumulate in specific sinks, such as soil, and river and ocean sediments, where they do not necessarily constitute an environmental threat.

DEFICIENCY, SUFFICIENCY AND TOXICITY

Many metals are nutritionally essential in small amounts, being necessary for plant, animal and human health and growth. Deficiencies may limit normal growth, development and reproduction, at times leading to visible clinical symptoms in plants, animals and humans.

Copper, iron, manganese, molybdenum, nickel and zinc are required by plants; copper, cobalt, iron, manganese, molybdenum, selenium and zinc are required by animals and man. Essential roles for arsenic, nickel, tin and vanadium have also been established in animal nutrition. Excessive amounts of some of the same metals may be toxic to plant and/animal life or may affect the quality of water or foodstuffs for human consumption. The concept of deficiency-sufficiency-toxicity is illustrated in Figure 1.

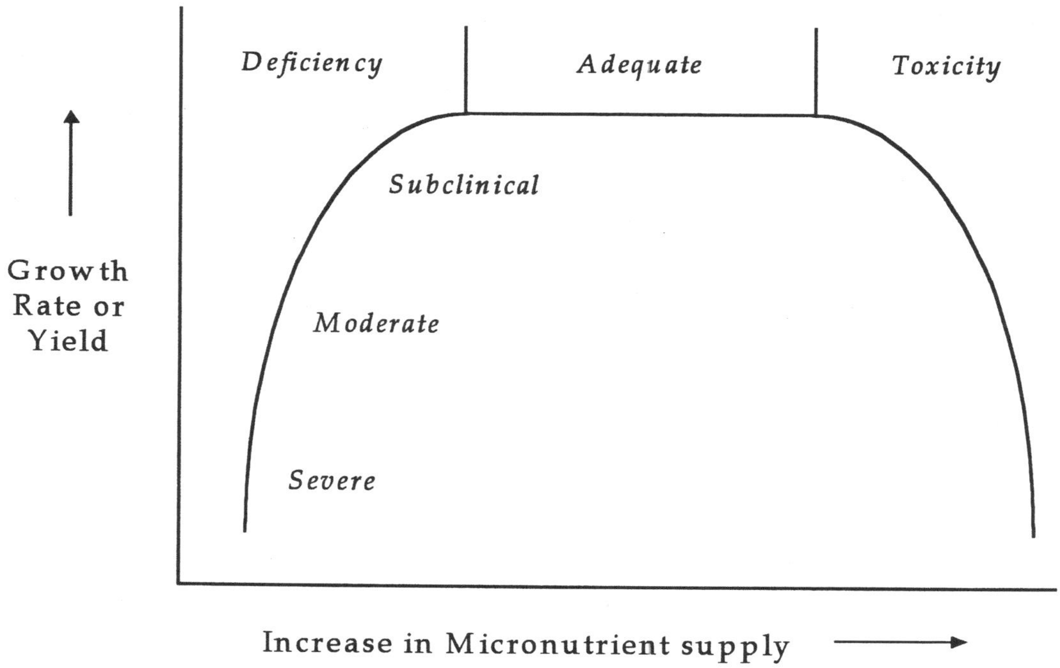

Figure 1. A typical Dose-Response Curve for Micronutrients such as Copper

Other metals, including lead, cadmium, antimony, thallium and mercury have no apparent nutritional or biochemical function. They are also present as natural components in the environment in small concentrations without giving rise to any toxic effects on life. However, if present in large enough quantities and in certain chemical or mineral forms, they may inhibit or impair plant and animal growth and health, or even lead to crop failure or death of farm animals. Metal toxicity in humans, however, is more often associated with occupational exposure.

The balance between the concentration (and form) of metals in the environment required to sustain plant, animal and human health, and levels at which toxicity may occur, is frequently difficult to quantify. For example, total concentrations of copper in agricultural soils below 5 μg/g are often associated with deficiency diseases in cereal and other crops, while levels of several hundred μg/g may be toxic to the same plants.

GEOCHEMICAL MAPS

Spatial information on the distribution of metals in the terrestrial environment on both a regional and a national scale has been compiled through geochemical mapping and the publication of geochemical atlases. These maps have been based on the systematic sampling and analysis of rocks, soils, stream sediments, lake sediments, surface waters and/or vegetation and were initially compiled for the purpose of mineral exploration (Hawkes and Webb, 1962). Webb (1964) introduced the concept of multipurpose geochemical surveys and the applications of geochemical maps to agriculture, pollution studies and public health. The prototype Wolfson Geochemical Atlas of England and Wales (Webb et al, 1978) was the forerunner of further geochemical atlases compiled by the British Geological Survey (Plant and Thornton, 1985) in the UK, and the many geochemical atlases that continue to be published elsewhere. Geochemical maps of Western Europe (Demetriades et al, 1990), and indeed a primary Global Geochemical Reference Database, under IGBP Project 259, based on wide-spaced sampling over the entire land surface (Darnley et al, 1995) are being planned.

Geochemical maps and atlases provide a useful time-dependent catalogue of baseline data showing the "natural distribution" of chemical elements, including metals relating to the composition of rocks and soils, and patterns of metal enhancement relating to mineralisation. Such information is of great help to those assessing the degree and extent of environmental pollution arising from human activities. They also provide primary reconnaissance data indicating areas in which nutritional deficiencies are likely to be found in agricultural crops, livestock and wildlife (Thornton, 1983)

URBAN GEOCHEMISTRY

Pioneering research in urban geochemistry has concerned the identification of metal sources in towns and cities and the degree to which they change or mask natural geochemical signatures in the surface environment (Kelly et al, 1996). Richmond upon Thames, one of the 32 boroughs within the Greater London area, was chosen as a residential suburb with little history of industrial activity. Systematic sampling and analysis of surface soils over an area of 56 km^2 has clearly shown a difference in the distribution of lead between those parts affected by urban development and the remaining areas of parkland and open space. The sites which show the highest concentrations of lead generally occur close to road junctions or roundabouts on main roads with high traffic density. Although unleaded petrol now accounts for over 80% of sales, a legacy of contamination from past years will remain far into the future, as lead has a very long residence time in soil. Parallel studies have shown that the use of platinum compounds in catalytic converters has resulted in detectable concentrations of this metal on heavily trafficked roads (Farago et al, 1994)

A comparative study of the city of Wolverhampton, with a long history of industrial activity, has shown contamination of zinc, iron and, to a lesser degree, cadmium and lead, in the vicinity of old industrial sites. Agricultural land on the outskirts of the city has significantly lower concentrations of all heavy metals studied. This study has clearly demonstrated that heavy metal concentrations in urban soils are strongly influenced by past industrial land use and must be taken into account in planning future development.

CHEMICAL AND MINERAL FORMS OF METALS

The complex processes involved in the formation and weathering of rocks and minerals has caused metals to exist in a wide variety of chemical and mineral forms. This is especially true for the surface environment. Some chemical forms of metals are, or can become, harmful to living organisms; however, many metal compounds are insoluble, chemically stable or not bioavailable. Others are demonstrably beneficial.

Very little of the total metal content of soils and sediments is present in free ionic or chelated forms in solution. Large amounts of metallic species are absorbed on the negatively charged surfaces of clay minerals and organic matter from which they can be removed by ion-exchange with hydrogen ions (from acids) and other cationic species. This labile or exchangeable fraction of the total metal inventory is largely bioavailable. Metals included in this fraction typically enter the food-chain by way of plants, sometimes aided by soil-dwelling micro-organisms.

Metals are also occluded in precipitated oxides of iron and manganese, are present as carbonates, or bound with insoluble species. Metal atoms may also be present in the lattices of secondary minerals. In these forms, metals are largely unavailable to plants and animals. Chemical species in which metals exist in primary minerals and many secondary minerals are relatively insoluble and of low bioavailability.

The implications of these different forms of metals in the surface environment with greatly differing states of solubility, mobility and bioavailability are of paramount importance in terms of the potential for metals to pass into the food-chain, act as essential nutrients, migrate into groundwaters, and pose a potential hazard to plants, animals and man.

BEHAVIOUR OF METALS

The behaviour of metals in soils is complex, and subject to a variety of soil properties and processes (Alloway, 1990). In this respect, it is stressed that soil is a dynamic system, comprising a variable mixture of mineral, organic and biological materials.

Soil pH is the most important factor controlling the chemical behaviour of metals in soils. Metallic cations such as Zn^{++}, Pb^{++} and Cd^{++} are most mobile under acid conditions, and an increase in pH reduces their solubility, mobility and bioavailability. However, some metallic anions such as molybdate and selenate become more soluble and available with increasing pH.

Another important process affecting the behaviour and bioavailability of metals in soils is that concerned with the absorption from solution onto solid phases in the soil matrix. The mechanisms involved include cation/anion exchange, specific adsorption, organic complexation and co-precipitation. Metal solubility and bioavailability is also influenced by the degree of soil wetness, organic matter content and type, and the interaction between one metal ion and others.

Geochemical studies of metal mobility at historically contaminated sites serve as a useful example of factors controlling metal migration through the soil-rock column. At three historical lead smelters in the UK, ranging in age from ca. 600 to ca. 1800 years, only small amounts of lead and zinc present in slag materials and contaminated surface soils appear to be in readily mobile forms (Maskall et al, 1995). It has been suggested that the metals are occluded in slag particles, precipitated as insoluble compounds, or are specifically adsorbed to soil constituents. The low mobility of lead in these soils has also been related to the elevation of soil pH by the release of calcium compounds from the slag waste during weathering. The rate of vertical migration of lead also varied with rock type, movement through sandstone being greater than limestone, which in turn was greater than a clay/sandstone matrix. The mean migration rates for lead and zinc showed a maxima of 0.75 and 0.646 cm per year (Maskall et al, 1995, 1996).

Concentrations of metals in air are strongly influenced by the proximity of local sources of emissions, since atmospheric dispersal processes lead to fairly rapid dilution. Particle size strongly affects atmospheric residence times; the smaller the particle, the longer it will remain aloft. Most metal-bearing aerosol particles range from 0.01-10 μm in diameter (Winchester, 1980). Particles larger than 10 μm settle out rapidly under gravity, while those smaller than 10 μm tend to remain in the atmosphere for longer periods, perhaps 10 to 30 days. Small particles can be transported as much as 400 km/day under moderate wind conditions. Wind-borne dispersion may result in long-range transport of metals currently being addressed by the UNECE Convention on Long-Range Transboundary Air Pollution.

The chemical form of airborne metal particles influences both mobility and the degree to which compounds are biologically absorbed following inhalation. For example, lead emissions from vehicles are mainly in the form of lead bromochlorides (although the proportion of halogen-containing compounds decreases with time and distance transported). Lead from wind-blown ores is usually in the form of PbS and $PbSO_4$, and that from smelters also contains $PbO.PbSO_4$ and PbO.

In general, metals entering the aqueous and marine environments react with other constituents in the water or with bottom sediments. This process largely results in their removal from solution and their fixation in the sediment and/or suspended particulate material. Many metals are adsorbed or co-precipitated with hydrated oxides of iron, manganese and aluminium. Thus, iron oxides co-precipitate vanadium, manganese, nickel, copper, zinc and molybdenum, while manganese oxides co-precipitate iron, cobalt, nickel, zinc and lead. The above processes may lead to the formation of secondary minerals in sediments, which tend to be insoluble or sparingly soluble. Natural geochemical cycling in rivers and oceans therefore tends to lead to a change of metals and their compounds from potentially bioavailable to less soluble forms.

GEOCHEMICAL ANALYSIS

In the 1990's, analytical techniques developed by geochemists have led to the development of methods for the identification and quantification of chemical and mineral forms of metals. This complex subject has been reviewed by Sposito (1983) who discusses the roles of chemical and physical factors that influence the presence of solid, adsorbed and aqueous species of metals in soils.

One approach to the analytical determination of the distribution of metals has been made by phase-selective chemical extractions involving single or multiple extracting reagents (Tessier et al, 1979). The reagents utilized in the sequential extraction have been chosen on the basis of their supposed selectivity and specificity towards particular physicochemical forms. It must be stressed, however, that these techniques are operationally defined and do not necessarily apportion metals into clear groups of chemical or mineral species.

An example is provided by the sequential extraction of soils from old lead mining and smelting sites in the UK (Li et al, 1995). With similar total lead concentrations, the partitioning patterns showed significant differences between the soils taken from a mining area and those from a nearby smelting area. The most important variation was the much higher exchangeable lead in the soils at the smelting sites, where the major contamination sources were emission particulates and slag. These sources contain anglesite ($PbSO_4$) and Pb-oxides (PbO and Pb_3O_4) which have a relatively high solubility in the $MgCl_2$ extraction step. Lead in soils in the old mining area is mainly in the form of cerussite ($PbCO_3$), galena (PbS) and pyromorphite ($Pb_5(PO_4)_3Cl$), which are of lower solubility in the $MgCl_2$ extraction step.

Analysis of individual particles of soil and dusts by scanning electron microscopy and x-ray analysis has been used to identify and characterize mineral species. For example, in an old lead mining area in Derbyshire, UK, the mineral pyromorphite ($Pb_5(PO_4)_3Cl$) has been shown to be the principal Pb bearing constituent (Cotter-Howells and Thornton, 1991). This Pb-phosphate mineral has a very low solubility and its abundance almost certainly contributes to the low bioavailability of Pb in these soils, limiting lead exposure in the local population.

RISK ASSESSMENT

Most risk assessment models on which regulatory guidelines and environmental legislation are based, require comprehensive site-specific data. In this respect and in many situations, it is clear that the role of the geochemist is of paramount importance in the underpinning of the risk assessment process with good science (Thornton, 1997). In particular, the geochemist can often provide valuable baseline data for naturally occurring concentrations of metals in soils and waters, maps showing regional patterns of metal distribution, information on the chemical and mineral forms of metals present in soils and sediments, and the influence of these forms on the solubility, behaviour and bioavailability of metals in the surface environment.

In conclusion, it is stressed that the bioavailable fraction of metals, whether as a source to plants and micro-organisms from the soil, or to animals and man through uptake from the gastro-intestinal tract, is frequently only a small proportion of the total concentration present. This is of paramount importance in relation to the supply of metals as essential nutrients or as potentially toxic substances. The development of appropriate methods for determining metal bioavailability, and the harmonization and wide acceptance of these methods both by the scientific community and regulatory agencies, is essential if risk assessment models are to be applied in the future in a more realistic fashion.

REFERENCES

ALLOWAY, B.J. 1990 Soil processes and the behaviour of metals. In Heavy Metals in Soils (ed. B.J. Alloway) pp. 7-28, Blackie, Glasgow and London.

COTTER-HOWELLS, J. and THORNTON, I., 1991 Sources and pathways of environmental lead to children in a Derbyshire mining village. *Environ. Geochem. Health*, 13, 127-135.

DARNLEY, A., 1995 A global geochemical database for environmental and resource management. Final Report of IGCP Project 259, UNESCO, Paris, France. 122 p.

DEMETRIADES, A., OTTENSEN, R.T. and LOCUTURA, K., EDITORS, 1990 Geochemical Mapping of Western Europe towards the Year 2000. Pilot Project Report: Geological Survey of Norway (NGU) Report 90-105.

FARAGO, M. et al. 1994 Platinum Group Metals in the Environment. Their Use in Vehicle Exhaust Catalysts and Implications for Human Health in the U.K. Report prepared for the United Kingdom Department of the Environment, June 1994.

HAWKES, H.E. and WEBB, J.S., 1962 Geochemistry in Mineral Exploration. Harper & Row, New York, Evanston and London, 415 p.

KELLY, J., THORNTON, I. and SIMPSON, P.R. 1996 Urban geochemistry: a study of the influence of anthropogenic activity on the heavy metal content of soils in traditionally industrial and non-industrial areas of Britain, *Applied Geochemistry*, 11, 363-370.

LI, X., COLES, B.J., RAMSEY, M.H. and THORNTON, I. 1995 Sequential extraction of soils for multielement analysis by ICP-AES, *Chemical Geology*, 124, 109-123.

MASKALL, J., WHITEHEAD, K. and THORNTON, I. 1995 Heavy metal migration in soils and rocks at historical smelting sites. *Environ. Geochem. Health 17*, 127-138.

MASKALL, J., WHITEHEAD, K., GEE, C. and THORNTON, I., 1996. Long-term migration of metals at historical smelting sites. *Applied Geochemistry*, 11, 43-51.

PLANT, J.A. and THORNTON, I. 1985 Geochemistry and Health in the United Kingdom. In Proceedings of the First International Symposium on Geochemistry and Health (ed. I. Thornton), pp. 1-15. *Science Reviews*, Northwood.

SPOSITO, G., 1983 The chemical forms of trace metals in soils. In Applied Environmental Geochemistry (ed. I. Thornton), pp. 123-170, Academic Press, London.

SWAINE, D.J. and MITCHELL, R.L., 1960 *J. Soil. Sci.*, 11, 347-368.

TESSIER, A., CAMPBELL, P.G.C., and BISSON, M. 1979 Sequential extraction procedure for speciation of particulate trace metals, *Anal. Chem.*, 51, 844-851.

THOMSON, M., THORNTON, I. and WEBB, J.S., 1972 Molybdenum in black shales and the incidence of bovine hypocrosis, *J. Sci. Food. Agric.*, 23, 871-891.

THORNTON, 1. 1983 Geochemistry applied to agriculture. In Applied Environmental Geochemistry (ed. I. Thornton) pp.231-266, Academic Press, London.

THORNTON, I. 1995 Metals in the Global Environment: Facts and Misconceptions. International Council on Metals and the Environment, Ottawa, Canada, 103p.

THORNTON, I. 1997 Risk Assessment Related to Metals: The Role of the Geochemist. Report of the International Workshop on Risk Assessment of Metals and their Inorganic Compounds. International Council on Metals and the Environment, Ottawa, Canada, pp. 133-140.

WEBB, J.S. 1964 Geochemistry and life, *New Scientist,* 23, 504-507.

WEBB, J.S., THORNTON, I., THOMSON, M., HOWARTH, R.J. and LOWENSTEIN, P.L. 1978 The Wolfson Geochemical Atlas of England and Wales, Oxford University Press.

WINCHESTER, J.W. 1980 Transport processes in air. In O. Hutzinger, editor, *Handbook of Environmental Chemistry,* pp. 19-30, Springer Verlag, Berlin.

One size doesn't fit all: Toward science-based risk assessment and management

George M. Gray, Ph.D.
Harvard Center for Risk Analysis
Harvard School of Public Health
718 Huntington Ave.
Boston, MA 02115 U.S.A.

ABSTRACT

A key understanding from the sciences of toxicology and epidemiology is that toxic substances are different. Different in the levels at which adverse effects are observed. Different in the way in which they are processed by the body. Different in the opportunities and pathways of exposure. Different in the mechanisms of toxicity. At the same time, current methods of risk assessment often treat all chemicals identically, ignoring knowledge from biology and chemistry. This quest for consistency in methods reduces the utility of risk assessments for risk management. Consideration of benefits or potential risk tradeoffs are confounded. Metals provide a clear example of the need to move beyond regimented, one-size-fits-all approaches to risk assessment and management to a system in which risks are characterized and managed by recognizing the specific and individual qualities of different toxic substances.

Proceedings of
Metals and the Environment: An International Symposium
Edited by Jim Skeaff
The Metallurgical Society of CIM
100th CIM Annual General Meeting, 1998

BACKGROUND

Risk assessment utilizes scientific information to estimate the potential harm from exposure to a toxic substance. The results of risk assessments then inform regulators, legislators, journalists, and the public about the magnitude of hazard they may face. Because there are many potential hazards it is important that risk assessment be able to distinguish the truly dangerous situations from those which pose little risk. It is scientific information, on potential human health effects, potential environmental damage, and sources and types of exposure, which should allow risks to be distinguished.

A general framework outlining the conduct of risk assessment has been developed by the United States' National Research Council, a committee of the National Academy of Sciences, in *Risk Assessment in the Federal Government: Managing the Process* (NRC, 1983). Many other governmental organizations, including the European Commission (EC, 1993), use a similar framework. The steps of risk assessment are described as hazard identification, dose-response evaluation (sometimes called effects assessment), exposure assessment, and risk characterization. The results of a risk assessment are then used to inform risk management decisions.

The evolution of regulatory risk assessment has tended toward a system in which consistency is a primary goal. In this way, no particular compound is treated differently than any other. Following the guidelines and checking off the boxes can come to replace scientific data and judgment in a quest for uniformity and consistency. As the use of the risk-based approach to management of potentially hazardous substances grows, it is crucial that the scientific distinctiveness of different compounds is recognized and considered.

Hazard identification is intended to determine the intrinsic dangerous properties of a substance for human or environmental health. Scientific information is examined to determine how strong the evidence is for an adverse effect. Weighing of evidence is necessary because substances of concern have very different degrees of information on the ability to cause harm. For example, with carcinogenic compounds hazard identification is supposed to weigh information from epidemiology, animal studies, short-term tests and mechanistic studies in order to characterize the carcinogenic hazard that a substances poses. The results of this analysis are often communicated to the risk manager, journalists, the public and others through the a carcinogen classification scheme (Graham, 1989). Both EPA and IARC have systems for evaluating and describing the evidence that a substance will cause cancer in humans.

Current classification schemes function in a checklist manner (is there sufficient evidence of carcinogenicity in humans? in animals?) and the boxes checked determine the classification. For example, if a compound has no human evidence and sufficient evidence in animals, it will be a class B_2 carcinogen in the EPA system. Mechanistic information indicating that the response in animals is unlikely to predict human risk does not change the fact that there is sufficient evidence of carcinogenicity in animals. In fact, it is difficult to envision how mechanistic information can influence the current system. In most cases, differences in hazard from different routes (*e.g.*, inhalation *versus* ingestion) or forms (*e.g.*, dust *versus* fumes) plays no role in hazard identification and classification.

In addition, critics of current approaches argue that reviewing scientists often give undue emphasis to positive results from long-term bioassays and do not incorporate mechanistic data, epidemiology, or negative test results, and hence current methods are not truly based on the weight of the evidence.

Dose-Response is the fundamental tenet of toxicology, exemplified by the oft quoted phrase from Paracelsus "All substances are poisons: there is none which is not a poison. The right dose differentiates a poison and a remedy." Once a hazard has been identified, the relationship between the dose of the agent and the probability of harm must be evaluated. The process of dose response evaluation evaluates the hazard associated with exposure to different amounts of a hazardous substance.

Determination of what dose of a chemical is safe and what is harmful has been the practice of toxicologists for many years, and the data from these dose-response relationships have been used in occupational health, environmental protection, and medicine to protect people against the toxic effects of chemicals. Central to this science of poisonous compounds and the determination of a safe level of exposure is the concept of a response threshold.

The threshold is the dose of the toxicant below which there will be no adverse effect. Above the threshold, some portion of the population will exhibit adverse effects. The dose-response relationship for a chemical is usually determined by tests on rodents, exposing them to a variety of doses of the compound and observing any toxic responses. The lowest dose producing a negative effect on the animals is called the lowest observed adverse effect level (LOAEL), and the next tested dose below the LOAEL is the no observed adverse effect level (NOAEL). The threshold dose in rodents, then, is somewhere between the LOAEL and the NOAEL, although its actual value is unknown.

When the rodent dose-response relationship is used to protect human health, the NOAEL is modified by safety (sometimes called uncertainty or assessment) factors, to account for potential differences in response between humans and rodents, to protect potentially sensitive portions of the human population, and to account for lack of knowledge of human response when there is little or no human data. Factors of 10 are generally used. For example, the NOAEL from a chronic animal test is generally divided by 100, a factor of 10 to account for potential differences between humans and the tested animal and a factor of 10 to protect potentially sensitive human groups. Factors of 10 (usually) are used to make other adjustments, from a less than lifetime test to a chronic RfD, from the LOAEL to the NOAEL if no NOAEL has been determined, if they are judged necessary

Factors of 10 have no particular scientific validity. It is rarely known how much more sensitive humans are than animals. Similarly, it is unknown what part of the human population may be more sensitive to a particular substance. In fact it is sometimes unclear that there is a sensitive population. It is apparent that the size of the factor necessary differs between chemicals and toxic effects. For some compounds and test conditions, the factor of 10 adjustment from animals to humans is a serious overestimate of the difference between the species. For others, the factor of 10 may even be too small. Current regimented approaches ignore differences in scientific data in the pursuit of consistency.

Dose-response evaluation for carcinogens in the United States differs from traditional toxicology because it is typically assumed that there is no threshold dose for a carcinogenic response. The linear no-threshold model, which is prominently used in cancer risk assessment, postulates that cancer can arise from a single change to the DNA of a single cell in the body. In other words, theoretically, a single molecule of a carcinogen has a nonzero probability of causing cancer. For this reason, it is assumed in cancer risk assessment that any dose of a carcinogen, however small, increases the probability of tumor formation (Anderson, *et al.*, 1983). Other groups, for example the United Kingdom Department of Health (1991), the European Commission (1993) , and the Health Council of the Netherlands, divide potential carcinogens into genotoxic and non-genotoxic categories. Dose-response evaluation for nongenotoxic carcinogens utilizes the NOAEL and Safety Factor approach while genotoxic carcinogens are assessed with a linear, no-threshold approach.

The quest for consistency in carcinogen risk assessment arises is manifest in the choice of mathematical models that is used to derive a dose-response relationship and extrapolate high-dose responses to low doses. There are several well-known statistical models for fitting the animal data and extrapolating the dose-response relationship to low doses. Often each model will fit the experimental animal data quite well and have at least some plausible basis in biology. The models nonetheless may yield low-dose risk estimates for the same chemical, from the same data set, that vary enormously (i.e., by factors of hundreds or even thousands). Yet the biological basis of the models is more appropriate for some compounds than for others. The default, conservative, methods of risk assessment used by the U.S. EPA assume a dose-response function that is linear in the low-dose region, has no threshold, and is proportional to the response at high doses. There is evidence that some agents, like certain types of radiation and directly mutagenic chemicals, may indeed have this type of dose-response relationship. However, many scientists believe the linear, no-threshold, approach to risk estimation is inappropriate for many other chemicals, such as those that are not direct mutagens. The countries and agencies that make a distinction between genotoxic and nongenotoxic compounds address part of this concern. However, even within the group of genotoxic carcinogens the appropriate dose-response model will vary between compounds.

Exposure assessment is the phase of a risk assessment that determines just how much of a potential hazard people or the environment are actually exposed to. Exposure to humans can occur through different routes, including inhalation, dermal absorption, and ingestion in food or water. In some situations, a pollution source will cause human exposure to occur through more than one exposure route (pathway), although multiple pathways are not always considered in EPA risk assessments. Environmental exposure can occur through accidental release, emissions during manufacture, use, or disposal of the substance.

When estimating exposure, data on the actual levels of contaminant in the relevant medium, along with ingestion rates would be very useful. The pharmacokinetics of the material in the body is a key link between exposure and toxicity. Yet actual data are rarely available. In some cases, such as assessing the risk of a facility not yet constructed, they are impossible to develop.

In the absence of hard data, the exposure assessment process proceeds with common "default" assumptions. For a pollutant found in drinking water, the U.S. EPA recommends that risk assessors assume that all people in the population consume 2 liters of water per day; another frequent assumption is that everyone in the population breathes 20 m^3/day of air. Additional important assumptions in exposure assessment include that 100% of a substance ingested in food or water is absorbed; that it is appropriate to average intermittent exposures over a lifetime; and that current levels of a contaminant in the environment will remain unchanged for 70 years. Food intake is generally derived from market basket surveys or national consumption surveys which gauge the amount of all types of groceries purchased or consumed. These assumptions used in exposure assessment usually do not take into account the heterogeneity of the population, including differences between the sexes, adults and children, different age groups, different ethnic and socioeconomic groups, and different levels of activity. Hence, exposure assessment is yet another source of uncertainty in the risk assessment process, sometimes leading to large overestimates of exposure, as in the MEI, or possible serious underestimates.

It is primarily the way in which numerical estimates of risk are characterized and communicated that has led to current dissatisfaction with the risk characterization process. Critics believe that although most risk characterization guidelines guidelines stress the need to address uncertainty in risk estimates and to emphasize the conservative upper bound nature of the risk estimates the guidance is routinely ignored. The degree of scientific validity in the choices of data, models, and exposure assumptions is not described. Instead, procedures designed to generate upper bounds on the risk are routinely treated as generating best estimates, and rarely is the uncertainty in risk assessment acknowledged. The biggest challenge for risk characterization is to move beyond the one-size-fits-all approach to risk to a system in which data and expert judgment allow the important distinctions between chemicals to be expressed and considered.

METALS - A KEY CHALLENGE FOR RISK ASSESSMENT

Metals provide challenges to the one-size-fits-all approach to risk assessment at every point; hazard identification, dose-response evaluation, and exposure assessment. Only through careful, scientific assessment can the risks of metals be properly assessed to allow sound risk management.

Several non-ferrous metals illustrate shortcomings in current practice of hazard identification. Nickel and nickel compounds are listed as known or probable (depending on the form) carcinogens by the U.S. EPA based on high dose exposures to animals and humans in the workplace. However, nickel may not pose a carcinogenic risk at lower, environmental, exposures. In fact, there is accumulating evidence that nickel in an essential trace nutrient which means that *lack* of nickel may be associated with health hazards. And, although the U.S. EPA lists nickel as a potential human carcinogen, the U.S. Food and Drug Administration (FDA) calls nickel "generally recognized as safe" (GRAS) and allows its use as a direct human food additive. Similarly, lead compounds, cadmium compounds, and beryllium and some beryllium compounds are called potential carcinogens based on high dose exposure with no consideration of whether it is scientifically appropriate to expect the same response from much lower doses. Zinc is considered toxic by the EPA based on high dose experiments while the FDA recognizes it as an essential nutrient.

In the context of non-ferrous metals several important issues arise in dose-response evaluation or noncancer effects. One of the most prominent is the essentiality of several metals for proper nutrition. For example, zinc in an essential component of more than 200 enzymes in the human body and, therefore, is necessary for normal function. The FDA has established recommended dietary allowances (RDAs) for zinc. Standard procedures for establishing an acceptable exposure level for zinc, based on factors of 10, lead to values close to, or below, the RDA. This illustrates the problem with using a regimented approach to all substances and not allowing alternative methods for characterizing the risk.

Another problem, although not unique to nonferrous metals, is the definition of a toxic effect. For example, the toxic effects of metals have been widely studied for many years. This means that there is a very large data base from which to draw in risk assessment. It is known that non-ferrous metals often have toxicological effects by mimicking or replacing other necessary minerals in the body. Therefore, exposures which result in systemic or organ toxicity are frequently preceded by a number of enzymatic or biochemical changes of questionable clinical significance. The problem is that these markers, rather than the toxic effect that occurs at higher doses, sometimes end up serving as the basis for risk assessment. These risk characterizations are then frequently placed on the same footing as those for other compounds derived from more serious toxic effects. This is a general problem with noncancer risk characterization.

A property of metals with important, but frequently overlooked, implications for assessing carcinogenic hazard is speciation. Metals can be found in many chemical forms, as salts, free ions, organometal compounds, and in different oxidation states. The form of the metal can have very important implications for toxicity and exposure. An example from carcinogen risk assessment is chromium in which hexavalent chromium (Cr^{+6}) is considerably more toxic than trivalent (Cr^{+3}) and only hexavalent are considered potential carcinogens. Yet many lists of carcinogens fail to distinguish between different forms of metals, some with clear carcinogenic risks, others judged guilty only by association. The form of a metal may also have important implications for pharmacokinetics and the dose-response relationship, requiring different approaches to risk assessment for different metal species. Clearly, a single approach is insufficient to capture the important scientific distinctions between metals and their many forms.

Exposure assessment for non-ferrous metals presents some situations which challenge standard default assumptions. Perhaps the most important issue is that of bioavailability. Bioavailability refers to the availability of a contaminant in some medium to be absorbed into biological systems. Different forms of metals frequently differ in bioavailability. A clear example comes from the case of lead in soil. Standard methods of exposure assessment generally assume that all lead in soil is bioavailable and able to be absorbed. Evidence indicates, however, that much lead in soil in a highly insoluble form and is not absorbed upon ingestion of soil. However, lead in other media or circumstances may be highly bioavailable. In assessing exposure, sound practice requires careful consideration of the form and milieu of metals. Scientifically complete exposure assessment must account for differences in bioavailability of different forms of metals.

THE NEED FOR SCIENCE IN RISK MANAGEMENT

The growing acceptance of the principles of risk assessment has increased its use in risk management. Many of the new uses of risk assessment involve making comparisons of health risks. Comparative risk projects have been undertaken to help states, localities, and even regulatory agencies rank sources of risk to citizens. Risk-based priority setting uses the results of risk ranking efforts to allocate resources for protecting health, attacking the worst problems first. Risk communication benefits from comparison between different types of risks to life so the public can understand the relative size of the risks of accidents, diseases, and environmental threats. Industry is beginning to compare different pollution prevention alternatives to find which generate the greatest reduction in risk. Finally, there is a growing emphasis, both nationally and internationally, on comparisons of the benefits (in terms of risk reduction) and costs of environmental, health and safety regulations.

Sound management of risk requires the best, science-based, estimates of potential harm. Methods that ignore the biological and chemical distinctions between substances distort risk estimates and confound efforts at benefit-cost analysis, comparative risk, and rational priority setting. As more and more is asked of risk assessment the practice must evolve. Risk assessors can no longer ignore basic scientific research about different mechanisms of cancer formation or disease development for different chemicals. Risk assessors can no longer be satisfied with worst case estimates of risk. Comparing risks will require best estimates of risk using the best science. To do otherwise is to invite misleading and incorrect comparisons. We must move beyond the one-size-fits-all approach to risk assessment to truly use all available scientific information.

REFERENCES

National Academy of Sciences, National Research Council, 1983 *Risk Assessment in the Federal Government: Managing the Process* , National Academy Press, Washington, D.C.

European Commission, 1993 Directive 93/67/EEC laying down the principles for assessment of risks to man and the environment of substances notified in accordance with Council Directive 67/548/EEC

Graham, J.D., 1989 Communicating about Chemical Hazards, Journal of Policy Analysis and Management 8:307-313

Anderson E. and the Carcinogenic Assessment Group of the Environmental Protection Agency, 1983 Quantitative Approaches in Use to Assess Cancer Risk, Risk Analysis 3:277-295.

UK Department of Health, 1991 Guidance for the Evaluation of Chemicals for Carcinogenicity. Committee on Carcinogenicity of Chemicals in Food, Consumer Products, and the Environment.

The socio-economic value of metals

Michael E. Henstock, Ph.D.
Consultant to the Reclamation Industries,
3, Brewery Yard, Sudborough, Kettering NN14 3BT, England.

ABSTRACT

Metals have been indispensable to humans since copper was first used some 9,000 years ago. They are prized for that blend of strength, hardness and heat resistance which gives them durability, and for their ease of shaping into complex forms. Improved technology has progressively reduced the price and improved the quality of metals, making them almost universally available and leading to improved living standards for countless people.

Metals are used by every sector of society, and for a widening range of purposes. Basic industry critically depends on them for machinery and plant. No aircraft, automobile, computer, or electrical appliance can function without them. Aluminium packaging reduces losses in food storage and transport, and aluminium sailing boats provide a less-expensive alternative to traditional wooden craft. Chromium is essential both to the stainless steel, which provides hygienic work tops and utensils in modern kitchens and is also at the heart of the chemicals industry, and of many of the alloys on which commercial aviation depends. Substitutes, when they exist, are often more expensive than metals. For many purposes metals are incapable of substitution and are therefore irreplaceable; substitutes, when they exist, are often more expensive than metals.

Proceedings of
Metals and the Environment: An International Symposium
Edited by Jim Skeaff
The Metallurgical Society of CIM
100th CIM Annual General Meeting, 1998

INTRODUCTION

Materials underlie the wealth and the standard of living of any society. No society can do without them. Many, if not all, major technological developments have arisen from improvements in materials or in the methods of processing them, or have had to wait for such improvements before they could be realised or attain full potential.[1]

Current standards of comfort and well-being in Western civilisation depend on wealth, the creation of which is critically dependent on the knowledge and use of materials for the manufacture of useful objects. The earliest materials, such as bone, hide, mud-brick, straw, stone and wood, have limited mechanical properties, and the difficulty of shaping them restricts their use. Old Testament references indicate that reliance on materials came in antiquity. Tubal-cain, we learn, was 'instructor of every artificer in brass and iron'.[2] Early attempts to use metals, which were doubtless discovered by accident, utilised their hardness and their ability to be hammered into shapes. From these activities came more complex operations: as humans discovered how to build fires hot enough to melt metals they also learned how to pour the melt into shaped cavities in clay or stone (**'casting'**), and thus to take up the desired shape.[3]

The first metal to be used was probably copper, which has only a limited ability to combine with other elements and is therefore sometimes found in the metallic state ('native' copper) in various parts of the world. Iron is more reactive and, although sometimes found in the form of meteorites from which early societies probably obtained their supplies, generally occurs combined with other elements, principally oxygen: thus, its widespread extraction and use came later than for copper. Two metals with pronounced resistance to atmospheric attack are silver and gold: as such, they are termed 'noble' metals. Their rarity, density, attractive appearance and ability to be melted and remelted from one form into another without loss of value led to their use in jewellery and coinage. Their possession became a measure of wealth. All other metals then known were, by contrast, termed 'base' metals.

Of the 92 naturally-occurring elements some 70 are metals. The properties of metals have made them almost indispensable to human endeavour since copper was first used some 9,000 years ago. Their blend of strength, hardness and heat-resistance gives them durability and their ease of shaping into complex forms enabled them to replace wood and stone in basic applications such as weapons and farm implements. Since early extraction techniques could not remove impurities the more-or-less systematic study of alloy formulation (**'alloying'**) followed almost by default. The first such alloy to be used was probably one of copper and tin (bronze), whose properties were found to depend on the tin content and, as the art of alloying developed, it was realised that the properties of all metals were markedly dependent on composition.

Other metals were valued not for their utility but for their value, as a function of their rarity. Gold, combining beauty, scarcity and permanence, still underlies currencies (though not physical coinage), but improving technology has reduced the cost of other metals, whose enhanced quality and availability have led to improved living standards for countless people. Indeed, the development of society since the Industrial Revolution has gone hand in hand with improved extraction methods for metals, whose history is a fascinating mixture of economics, science, superstition and technology.

Metals are used by every sector of society. Basic industry critically depends on them for machinery and plant, and industrialisation, the source of wealth in developed countries and the goal of every developing society, has the same dependence. Potential substitutes are often more expensive than metals. Life-Cycle Analysis, which takes into account the durability and long life of metals, may also show that they are economical in terms of the energy of extraction and of conversion into consumer goods.

No aircraft, automobile, computer, or electrical appliance can function without metals. Iron and steel are still the sinews of industry, for automobiles, ships and trains. Zinc protects those steel structures from corrosion, whether the automobile body from the salt spread on roads in winter or the oil rig from the ravages of sea water.

Electrical power, generated and supplied using copper conductors, makes 20th Century life, as we know it, utterly dependent on copper; aluminium is the only feasible substitute. Copper windings are at the heart of every electric motor that, housed in complex, accurate and cheap metal die-castings, operates food-blenders, CD players, dishwashers and photocopiers. Cadmium, calcium, lead, manganese and nickel drive the batteries that provide portable power for automobiles, calculators and radios. Aluminium packaging reduces losses in food storage and transport, and aluminium sailing boats provide a less-expensive alternative to traditional wooden craft. Chromium is essential to the stainless steel which provides hygienic work tops and utensils in modern kitchens and which is also at the heart of the chemicals industry, and to many of the alloys on which commercial aviation depends.

Rare metals, with an annual production measured in the range of 100 tonnes or less (compared with the 20 Mt of aluminium and 10 Mt of copper) provide the magnets for advanced loudspeakers and are critical components of the semiconductors which are essential to the inexpensive electronic goods produced in such profusion in the countries of the Pacific Rim. The list is literally endless.

Advanced societies find metals indispensable for an increasing range of purposes. Their use continues to evolve: only in this century has aluminium been transformed from laboratory curiosity to the most widely-used non-ferrous metal; it plays a major part in metal packaging which, in 1991, was estimated to account for some 23% of the total packaging market in the United Kingdom, with an estimated gross market value of just over £2,000 million ($3,000 million).[4] Only since World War II has titanium been available for use in the supersonic aircraft that rely on it to withstand aerodynamic heating, and the list of so-called 'new' metals may also be extended to include niobium ('columbium' in the USA), tantalum, zirconium, and the so-called 'rare-earth metals'.

For many purposes metals are incapable of substitution: they are literally irreplaceable, especially when cost per unit of useful property is taken into account. However, few occur naturally in usable form. Historically, they were rare and precious commodities, available only with great effort and expense. Population growth and increasing individual appetites for goods have increased the demand for them. This has been met in part by increased production, in part by their substitution (notably by plastics), and in part by the expansion of another ancient activity, recycling.

Financial necessities and continued and increasing pressure to reduce the impact of manufacturing processes and their products on the quality of life seem likely to encourage humanity to do more with less material. As examples, locomotives that weighed 25 kg/hp in 1950 weighed only 15 kg/hp in 1980, while between 1975 and 1985 there was a 20% reduction in the weight of an aluminium beverage can.

Concern continues regarding possible shortages of metals. Since their geographical distribution is not uniform it is probably true that no one country can supply from within its own borders all the materials required for its continued development. This irregular distribution of ores makes some metals of critical (**'critical metals'**) or strategic (**'strategic metals'**) importance. Critical metals are generally taken to be those with a high degree of need in the defence industry. Strategic metals are those which, in addition to being critical for the aerospace and directly associated industries, also have a marked degree of import dependence and a limited number of significant suppliers.

Such a combination confers geopolitical vulnerability. Chromium, which is heavily concentrated in southern Africa, is a notable example of both categories, but NATO and other organisations have, at one time or another, identified cadmium, cobalt, copper, lead, silver and tungsten as likely to fall short of world requirements. This could lead to such metals becoming increasingly expensive and being substituted or limited to specific applications. However, these gloomy forecasts have yet to be realised.

Their ability to be recycled has always made metals attractive. Before the Industrial Revolution it would have been almost unthinkable to discard metal objects. Clearly, the prophet Isaiah was describing a well-established and profitable practice when he wrote: '... and they shall beat their swords into ploughshares, and their spears into pruning-hooks'.[5] Today, recycling is a great international industry, with a world-wide value (1989) of some $16,000 million, while in 1994 it employed a workforce of approximately 100,000 in 15,000 businesses in the European Union (formerly the European Economic Community, or EEC) alone. In 1993, the total EEC production of the four major non-ferrous metals (Al, Cu, Pb, Zn) was 8.184 Mt. Of the 4.698 Mt of scrap consumed in the production of these tonnages, approximately 400,000t was imported from non-EEC countries.[6] Of total United States consumption in 1993, secondary metals provided 38% of the aluminium, 41% of the copper, and 66% of the lead. For Japan, the corresponding figures were: aluminium 35%, copper 36% and lead 24%. In contrast with the popular images of the 'rag and bone' type the industry is highly mechanised and closely regulated to conform with relevant environmental laws.

Traditionally, metal recycling is rooted in the economic laws of supply and demand. Its extent varies according to circumstance. Recycling levels increased considerably during the Great War and again during World War II. Prosperous peacetime demand may exceed the capacity of primary metals suppliers and existing scrap flows: the resulting price rises then bring additional scrap into the marketplace and make reclamation profitable from previously unattractive areas. Metals are not the only recyclable materials but they are especially well-suited to it. They have a high monetary value, and secondary and primary can be indistinguishable; thus, metals predominate among substances that are commonly reclaimed, and recovery can confer major economic and environmental advantages, especially for metals that are used in large tonnages, are exceptionally rare or expensive, or are toxic.

Scrap is an integral and often essential component of the metals supply system and of smelter economics, and is actively sought by the integrated metallurgical companies. In industrialised countries with mature economies, a steady supply of scrap enters melting furnaces from manufacturing and from goods reaching the end of their useful life. In 1995, Western countries consumed 10.347 Mt of refined copper, of which 1.691 Mt (16.3%, an increase over the 15.9% in 1993) came from scrap; a further 3.337 Mt of scrap was recovered and used without refining (**'direct use of scrap'**).[7] Thus, of total copper consumption in that year, 36.7% had already had one or more service lives.

Secondary metals command the same price as primary of equal purity. Less pure secondary metals are almost always cheaper per tonne to buy than primary, reflecting both some lack of flexibility in use and the cost of refining them to primary specification. However, secondary may often be used without refining, offering cost savings: in the United States, 67% of all copper scrap consumed in 1995 was used in this way, providing 28% of total copper consumed there.

Recycling extends reserve lifetime and reduces solid waste (litter), mining wastes and associated pollution, import dependence and demands on landfill. Under favourable conditions it can also reduce both overall energy consumption for metal production and the associated pollution of energy provision. After 1973, increased energy costs reduced growth rates for both primary and secondary metals, but the percentage reduction in primary aluminium consumption, for example, was three times that for secondary.[8]

Recycling has no virtue in itself. Generally, if it is financially profitable it is done. It is desirable only if it produces a net gain as judged by valid yardsticks. There may be no advantage if reclaimed material costs more than primary, or if the additional energy consumption or the emissions of collection and processing outweigh the savings.

No substance can be recycled completely: attempts to collect all recyclable material invariably increase both total energy consumption and the pollution from fuel combustion to an extent which can easily exceed any savings. Nevertheless, secondary aluminium can be produced from suitable scraps, such as engine blocks, for as little as 5% of the energy required to make primary, and secondary copper may consume only 3% of the energy associated with primary metal.

It is, however, accepted that recycling is often worth-while for other than financial reasons. The reclamation industries protect the environment, by processing residues which would otherwise be abandoned, from ashes to electronic equipment, and convert them into usable metal in an environmentally-sound manner, using methods that are often complex and capital-intensive. Their international trade makes the advantages of such methods available to countries which lack the relevant equipment.

Recycling can, however, never satisfy the increasing demands of an increasing population. Assuming an average product lifetime for copper of 30 years, complete recovery of the 7.132 Mt (5.027 Mt of refined metal plus 2.105 used directly) in Western countries in 1965 and notionally available for recycling in 1995 would fall far short of the 13.684 Mt (10.347 plus 3.337 Mt used directly) of the latter year.[9]

ALLOYING AND METAL PROPERTIES

Early in the history of metal use it was discovered that mixtures of metals, termed 'alloys', could exhibit hardness, strength or toughness superior to those of their components. An alloy may be defined as a material possessing metallic properties and comprising two or more elements, at least one of which is a metal. Alloying usually changes the properties of both. Bronze, an alloy of copper with around 10% tin, was an early discovery, and it soon became clear that its properties depended upon the relative proportions of the constituents: more tin increased the hardness. This revelation opened the way to further improvement of properties by so-called 'heat-treatment', whereby the alloy is heated to a selected temperature, below its melting point, and then subjected to some predetermined cooling and reheating régime to obtain specified properties.

Very few metals are used unalloyed. The choice of a metal or, more frequently, of an alloy for a particular application depends on its ability to meet a given set of criteria. Of the several such alloys that may meet the requirements, the cheapest is usually selected. The alloy with the lowest *purchase price*/tonne is not, however, always the cheapest to put into service: a more expensive material may be easier (and therefore cheaper) to work, or to have a lower wastage rate, or to yield scrap of higher value.

The properties for which metals or alloys are selected are many, and often interrelated.

Abrasion resistance: Abrasion resistance is resistance to surface scratching or deformation, a property closely allied to hardness (*q.v.*). Some alloys, for example steel or aluminium-copper alloys, may be hardened, at the surface, through the entire section, or both, and can therefore be made to have a hardened, scratch-resistant surface to a ductile ('tough') interior, so conferring superior mechanical properties.

Appearance: Copper, gold and their alloys have their own attractive self-colouring; the subtle rose-pink of electro-deposited tin-nickel alloys or the blue-white of tin-zinc are less well known. The reactivity of aluminium and its alloys with oxygen results in the formation on their surface of a thin, tenacious oxide film, which may be thickened electrolytically and then dyed to almost any desired colour.

Castability: Casting involves pouring molten metal into a mould, often of complex shape, to produce a solid object in that shape. Examination of an unweathered bronze statue or of a die-cast model car reveals the notable ability of metals to take up detail.

Catalytic properties: A catalyst accelerates, or even makes possible, a chemical reaction while itself remaining unchanged. Well-known examples include the use of vanadium pentoxide (V_2O_5) to convert sulphur dioxide to sulphur trioxide in the manufacture of sulphuric acid or, perhaps more familiarly (although very few motorists probably ever consider it), the use of platinum, palladium and rhodium in automotive catalytic converters, which reduce exhaust emissions from a car engine.

Conductivity: Both the electrical and the thermal conductivity of metals are generally at least an order of magnitude greater than that of almost all non-metals. The generation, distribution and use of electricity utilise conduction, almost always by copper, whose conductivity exceeds that of any other metal except silver. Today, aluminium is often used: it has 61% of copper's conductivity, necessitating thicker sections for equivalent current-carrying capacity, but its lightness (specific gravity less than one third that of copper) makes it attractive in long spans of overhead power cables, where it requires a steel core in order to withstand its own weight without stretching.

Heat transfer depends on the thermal conductivity of the barrier between hot and cold areas. Familiar examples include transfer from hot water inside a central-heating radiator to its external surface, or from the interior of a building by air-conditioning.

Corrosion resistance: Corrosion is the destructive attack of a material by chemical agents such as water (whether pure, so-called 'acid' rain, or seawater), contaminated air or aggressive chemicals. Though materials such as concrete also suffer from it, metals provide spectacular examples of corrosion. Iron and steel bridges, ships or motor vehicles are corroded to form the bright orange-brown corrosion product termed 'rust'. In the United Kingdom alone the annual bill for corrosion damage is estimated as $8,000-11,000 million. Corrosion occurs through high-temperature attack by elements such as oxygen (oxidation) or sulphur (sulphidation). However, suitable alloys, notably the iron-chromium-nickel alloys known as 'stainless steels', have superior corrosion resistance, and remain almost inert in all but the most severe environments.

Creep resistance: 'Creep' is a failure mode displayed under uninterrupted loads whereby metals yield slowly and, over a period which may be years, stretch and eventually fracture. Progressive improvements in the gas-turbine aircraft engine have required alloys that resist creep under ever-higher stresses and temperatures, in the corrosive environment of fuel combustion products. At a more everyday level, the elements in electric stoves and the tungsten filaments in light bulbs must possess high creep-resistance in order to continue to function at red or white heat.

Ductility: The ability of a material to undergo permanent deformation without fracture is termed 'ductility'. It is an essential quality in, for example, the mild steel that is still the most suitable material for shaping into the complex forms of automotive body parts.

Energy return factor (ERF): Energy return factor denotes energy savings that may be made *during the operating lifetime* of an object by the use, in its construction, of lightweight materials. An obvious example is the replacement of steel or cast iron (specific gravity approximately 7.8) by aluminium (specific gravity 2.7) or plastics (polyethylene 0.91-0.96, PVC 1.38) to reduce the weight of road vehicles in the interests of fuel economy. The fuel saving over a vehicle lifetime may then exceed the additional energy consumed in the production of the lighter materials. For a notional United States passenger car the ERF attainable by using aluminium alloys in specified components has been calculated as 2.7, despite the fact that the alloys may, for equivalent strength, need to be used in greater *volume* than the steel they replace.[10]

Fatigue resistance: Repeated stress reversal, especially under corrosive conditions, may cause failure at lower stresses than under steady loading. Fatigue is claimed to account for more than 90% of all mechanical fractures.[11] For a given component (the design is important) there is a certain stress, known as the 'endurance limit' or 'endurance strength' below which fatigue failure does not occur, no matter how many the reversals. Usually, 10 million cycles is termed 'infinite life' and is satisfactory for determining endurance limits.

Formability: Formability describes the ability to undergo mechanical working or shaping without rupture. A significant part of the factory-gate cost of any object arises in forming costs. It follows that the use of a more expensive material (in terms of cost/tonne) may be justified if its better formability simplifies and cheapens production.

Hardness: Hardness signifies resistance to deformation. Most metals and alloys may be hardened by mechanical work ('work hardening'), while some alloys are capable of being hardened by a régime of heat treatment.

Magnetic properties:. Many powerful magnetic alloys are based on nickel; examples include Alcomax (Ni-Fe-Al), Alnico (Al-Ni-Co-Fe-Cu), and Ni-78.5% Fe, the last of which is unusually susceptible to low-intensity magnetism. Cobalt-samarium magnets are a more recent development.

Strength: Strength is the term used generally to indicate resistance to applied stress of any form. In metals, it represents an ability to resist several types of loading without permanent change of shape ('plastic deformation').

Toughness: Toughness is defined as the work required to fracture unit volume of material. It is dependent on the conditions of testing, especially on the temperature and the rate of load-application. The use of unsuitable steel, which became brittle at Arctic temperatures, led to the loss of several Liberty Ships during World War II.

COMPLEX CRITERIA FOR METAL SELECTION

Only occasionally is it possible to specify a material for one criterion only. A requirement for strength can rarely ignore weight: thus, the concept of strength-to-weight ratio is important. Titanium has a yield strength of 415 MPa (60,000 psi), which can be raised to 1,300 MPa (190,000 psi) by alloying, comparable to that of many alloy steels, yet with only 56% of the density of steel. Its strength-to-weight ratio makes it attractive for use in high value applications, for example aerospace, where its cost can be accepted provided that service temperatures do not exceed 480°C.

Since cost is critical the concept may be extended to cost or, generally, cost/unit of property/unit of weight. Much of the world tonnage use of copper is in electrical conductors: calculated *solely* on conductivity and cost/tonne, the cost of copper per unit of conductivity is more than three times that of aluminium. This is not, of course, the only relevant criterion in choosing materials for electrical conduction.[12]

The success of technology in improving the properties and reducing the cost of materials gives the engineers a greater choice than ever before. Materials such as aluminium, plastics, steel and wood are in constant competition. Enormous sums of money are invested by the producers of primary materials in an effort to win market share. The outcome of their efforts depends on decisions taken by engineers who design new products or redesign old ones. The knowledge that there is still potential for increased commercial uses of metals, especially the newer ones, has prompted industry also to invest in teaching and research, with emphasis on their particular product.[13]

METALS AVAILABILITY

Some metals, such as aluminium, are widely distributed, although only relatively few areas contain commercially workable concentrations of Al_2O_3. By contrast, known deposits of chromium - so far incapable of substitution in the highly creep-resistant alloys used in modern gas-turbine engines - are heavily concentrated in southern Africa. Some 95.4% of total known reserves are found in Zimbabwe and the Republic of South Africa, which goes far toward explaining the difficulty, when such things were attempted, of applying economic sanctions to those countries.[14]

The consumption of metals has grown throughout the 20th Century. More metal has been extracted since 1900 than in the thousands of years of previous recorded history. Table 1 gives the annual world production of some base metals in various years.

The economic value of an ore critically depends on the market price of the metal and on the ease of extraction. At the average 1994 gold price of approximately $389/troy oz. only

Table 1

PRODUCTION OF THE COMMON METALS

Metal	Average production per annum (Mt)				
	1858-1907	*1908-1957*	*1958-1970*	*1971-80*	*1980-1993*
Aluminium	0.002	0.77	6.40	13.66	17.65
Copper	0.26	1.42	5.75	8.68	10.15
Lead	0.06	1.49	3.15	4.83	5.31
Magnesium	0.0	0.03	0.15	0.27	0.32
Nickel	0.004	0.05	0.39	0.67	0.78
Steel	11	120	475	690	790.2 [15]
Tin	0.06	0.15	0.19	0.23	0.22
Zinc	0.30	1.41	3.94	5.81	6.50

Sources: Alexander & Street (1858-1907)
Metallstatistik 1983-1993 (other data)

mines working rich deposits or enjoying low local labour costs (principally in South Africa, Russia and Brazil) could operate profitably. At the average 1980 US price of $839/oz., old mines reopened and old mine tailings dumps were reworked with modern techniques capable of extracting gold left behind by earlier, cruder, methods.

Another factor in availability pertains to chemical affinity. The extraction of relatively inert metals, such as lead (39 GJ/t) from typical ores requires much less energy than does that of aluminium (270 GJ/t). The ability to extract metals depends, *inter alia*, on the availability and price of energy, with a major influence on the cost of the metal produced. Energy factors, taken with geological and geographical considerations, make it clear that metallic availability is only a relative concept.

Anxiety has been expressed concerning the probable lifetime of many metals and of certain non-metals, such as oil and gas. In 1970 Meadows *et al.* published their influential book *The Limits to Growth*, which indicated that, at the then levels of extraction, certain metals would be exhausted within lifetimes (termed 'static index') of as short as 9 years.[16] The analysis was then extended to consider the effects of increasing population and demand for minerals, to determine the probable lifetime ('exponential index') under conditions of increasing demand.

There have, however, been many other false alarms: for example, a NATO study (1976) suggested that a number of metals, including copper, lead, mercury, silver and tin would be scarce before the year 2000, but there is no evidence that that prediction is likely to be fulfilled. The term 'known global reserves' must be interpreted with caution. Such terms are, of necessity, hedged about with many caveats. Perhaps the most important is the linkage between availability and price. In effect, 'reserves' are the commercially useful portion of 'resources' (concentrations of naturally-occurring material from which economic extraction of a commodity is currently or potentially feasible) under prevailing economic conditions. Inasmuch that seawater contains gold (approximately 48 kg/km^3) it is a *possible* resource for that metal, but is plainly not a reserve.

If ores become leaner, more complex and more difficult to process the energy for extraction rises rapidly. The lowest usable ore grade is markedly dependent on the availability and price of energy to extract from it and on our progress toward containing the environmental effects of their exploitation, since the extraction of 1 t of metal from a 1% ore necessarily involves the handling of ten times as much material as from a 10% ore, with commensurate environmental impact. In 1975 it was estimated that to extract lead from common rock, where its concentration is 15 ppm (0.0015%), could cost up to $1 million/t. A vehicle battery would then cost $10,000.

It is noteworthy that all metals considered in *The Limits to Growth* continue to be available, in some cases years after the date of their predicted exhaustion. Moreover, since raw materials account for only some 4% of global gross national product, even a sharp rise in the prices of mineral raw materials would have only a small effect on the standard of living of much of humanity.[17] But it is also undeniable that primary metals are finite, albeit only in the long term, and that many of their uses are dissipative.

CONCLUSIONS

By the year 2000, world demand for electrical power may well be three times that of 1980. Though demand in some developed countries, such as Canada, has fallen significantly, the developing countries will require more power stations, generators, distribution lines, cables, switchgear, and everything else that goes with an increased population and with its appetite for domestic and other electrical appliances.

Metal demand in developing countries, whose *per capita* use has previously been small, may then be much greater. The *per capita* consumption of non-ferrous metals by an enlarged consuming population is, however, unlikely ever to reach the levels already reached in the developed world. Even assuming that the future wealth of the developing countries gives them comparable purchasing power, future unit consumption is certain to benefit from the reduced materials requirements of modern designs.[18]

The epitaph that many were ready to pronounce for metals in the face of their rapid substitution by polymers and ceramics in some traditional applications has, like those for the death of Mark Twain, proved to be greatly exaggerated. At present growth rates, population is the principal cause for concern in their availability.

REFERENCES

1. Henstock, Michael E., *The Recycling of Non-Ferrous Metals*, Ottawa, International Council on Metals and Environment, (1996).

2. Genesis, iv, 22.

3. Alexander, William and Arthur Street, *Metals in the Service of Man*, (Seventh Edition), Harmondsworth, Penguin Books Ltd., (1979), pp.19-25.

4. Staff, Key Note Publications Ltd., *Market Sector Overview: Packaging (Metals & Aerosols)*, Eighth Edition (1991), p.2.

5. Isaiah, ii, 4.

6. Private communication, Bureau International de la Récupération, (June 23 1995).

7. *Metallstatistik 1985-1995*, (83rd Edition), Frankfurt am Main, Metallgesellschaft AG, & Ware, World Bureau of Metal Statistics, (1996). Secondary metal is not always used in the year in which it is produced from scrap.

8. Henstock, Michael E., 'The European Picture', Chapter in *The Impacts of Material Substitution on the Recyclability of Automobiles*, New York, The American Society of Mechanical Engineers (ASME), (1984), pp.171-185.

9. World consumption of refined copper in 1995 (11.803 Mt) was slightly lower than production (11.533 Mt); a further 3.337 Mt (for Western countries alone) of copper scrap was used without refining (direct use).

10. Cochran, C.Norman 'Aluminum - Villain or Hero in Energy Crisis?', *Automotive Engineering*, 81, (6), pp.57-61, (1973). The distance covered must be specified: Cochran assumes 160,000 km (100,000 miles).

11. DeGarmo *et al*, p.52.

12. Alexander, W.O., 'Comparative values of metals and materials', *The Metallurgist - and Materials Technologist*, (June 1973), pp.291-295. Alexander's data have been recalculated using an aluminium price of $1,455/t and a copper price of $2,424/t, as compared with his 1973 figures of £425 and £630/t respectively. The balance is, of course, dependent on the relative prices of the two metals at any given time.

13. Hydro Aluminium, for instance, has endowed a Chair in Industrial Design at the Norwegian Institute of Technology, see: Dag Flaa, 'Knowledge - a weapon in the battle between materials', *Aluminium*, 69, (8), (August 1993), p.673.

14. Papp, John F., 'Chromium', *Minerals Yearbook, Volume I, Metals and Materials*, Washington, D.C., United States Department of the Interior, Bureau of Mines, (1993), pp.281-2.

15. Average 1987-1991.

16. Meadows, Donella H. *et al, loc. cit.* It is not the purpose of the present work to examine long-term global models. Certain of the predictions of *The Limits to Growth*, a landmark in our perception of the global system, may now be seen to have been too pessimistic: these include the forecasts of the availability of metals. Nevertheless, despite continuing efforts to discredit the work, certain other conclusions now (1995) appear to have been uncomfortably accurate.

17. Callot, F., *loc. cit.*

18. Themelis, Nickolas J., 'Pyrometallurgy near the end of the 20th century', *JOM*, (August 1994), pp.51-57.

The Canadian "Metals in the Environment" research program

Bruce R. Conard
Inco Limited
145 King Street West
Toronto, Canada M5H 4B7

ABSTRACT

There are a number of countries and international organizations debating the criteria for the safe use of metals within mining, metallurgical, manufacturing, and customer use perspectives. The issue is to determine long term risks that metals may entail and to develop risk management strategies that will minimize adverse outcomes.

Canada is taking a leadership role in discussing what information may be needed to assist in assessing and managing environmental and health risks due to metals. The Mining Association of Canada and Canadian departments of Environment, Natural Resources, Health and Fisheries and Oceans, have formed a consortium called MITE (Metals in the Environment) to identify what is known and define what is needed to be known in order to use metals safely. Integrated, priority-driven research is envisioned over a 5-year interdisciplinary program. Liaison with other programs internationally will be a feature of this program.

The research will cover three principal areas, improving knowledge about:

(a) sources, that is, improving estimates of natural releases of metals into the environment and developing and validating methods to distinguish between anthropogenic and natural contributions;

(b) processes affecting fate, that is, identifying the processes that occur in atmospheric, geological, hydrospheric, biological and terrestrial compartments and which influence the transport, speciation and ultimate fate of metals;

(c) effects, that is, linking exposures of metal species to cellular, individual and population responses in both acute and chronic timeframes.

Details of the MITE research program are described.

Proceedings of
Metals and the Environment: An International Symposium
Edited by Jim Skeaff
The Metallurgical Society of CIM
100th CIM Annual General Meeting, 1998

INTRODUCTION

After several decades of political and regulatory pressure on environmental issues related to synthetic organic chemicals, the 1990s has been a decade where attention is being shifted to inorganic materials.[1,2] This attention will not diminish in the foreseeable future. It will grow and be joined by a new emphasis on risk assessment and risk management of metals.* But such efforts must be based on sound methods and reliable information.[3]

Most people support care for the natural environment, responsible use of resources and stewardship of industrial products, but the manner in which some governments and regulators are dealing with metals and metal products is causing considerable concern. There is the tendency, for example, for regulators to want to shoe-horn metals and metal compounds into the regulatory framework for organic chemicals. A question is asked: are metals biodegradable? When the answer is "No", metals are assumed to have a long-term danger for the environment and for living organisms. People who ask such questions have ignored that the earth is composed of minerals and, as such, metal compounds are ubiquitous in our natural environment. A factor which also is often ignored is the essentiality of a large number of metals to living organisms.[4] Organisms must have ready access to essential nutrients and they do so by bio-accumulating metals. Organisms have also developed processes, known collectively as homeostasis, to regulate the inventory. Such fundamental differences between metals and synthetic organics must be taken into account in writing regulations or else substantial harmful health effects to humans and ecosystems may unwittingly be caused.

A feature of the environmental scrutiny of metals is the assertion that, as pollutants (i.e. unwanted substances), they are generated by human activity. There is little doubt that this is the case for synthetic organics: humans synthesize them. There is doubt, however, about where metal "pollutants" come from. A portion certainly comes from human activities.[5] Increasingly, however, there are indications that a very large portion may come from natural sources.[6,7] The scientific debate is about the relative importance of each source. This debate is quite significant as to whether the management schemes conceived and applied will be successful. It would be tragic if human talent and resources were focused on controlling metal pollution by controlling human activities, only to realize afterwards that the natural environment itself was the overwhelming controlling factor.

Another feature of the issue of metals and their effects on the environment and living organisms is the paucity of reliable scientific data that are relevant to the long-term risk of chronic occurrence of metals in the environment. Thinking is occurring, positions are being taken, regulations are being developed and laws are being passed in what is often a scientific vacuum. The danger is that society will end up with misaligned priorities and with attitudes and legal frameworks that severely penalize the mineral resource industry and these conditions will jeopardize the industry's ability to contribute, in a responsible way, to human development and well-being.

---

* Throughout this paper, when the term "metals" is used, it encompasses all metal-containing inorganic substances.

THE CANADIAN PERSPECTIVE

The minerals and metals industry in Canada directly employs some 350,000 people in 150 communities and generates about 10% of the Gross Domestic Product. Canada is known world-wide for its peacekeeping, its health care, its multi-ethnicity, its forests, its water and its metals. Not only must Canada concern itself with sustaining its natural heritage, but it is looked to by the world community to provide leadership in environmental issues. Of the many strengths that Canada has are the skill, education and dedication of the human resources within its metals industry.

Canada's leadership in environmental and health activities at the United Nations, the Organization for Economic Cooperation and Development (OECD), the World Health Organization (WHO), etc. is evidence of the strong and influential role Canada wants and is expected to play. However, several Canadian ministries have conflicting views on some basic questions related to metals in the environment. Accordingly, Canada's message to the world community is sometimes confusing because representation is made by various sectors of Canadian government or industry. The net result is that Canada often may not have as effective a voice as it should.

The trend toward international treaties and agreements in environmental affairs is gaining momentum. More and more pollution abatement issues are being discussed because of the trans-country movement of water and air. Most of these treaties require that signatories promulgate laws that specifically enforce the treaty. Thus, we will find that more and more of Canada's national environmental laws will arise from activities in the international arena. Such events make it even more essential that Canada have an integrated perspective on the safe use of metals and that Canadian representatives contribute sound scientific information for use in the international arena, as well as in our nation's laws and policies.

THE MINING ASSOCIATION OF CANADA (MAC)

Worried by anti-metals activities of environmental activists and some governments, recognizing that Canada has an important role to play in fostering responsible and sustainable use of metals, and concerned that scientific harmony was not present among leading stakeholders in Canada, MAC sponsored a workshop at Val Morin, Quebec in October, 1996, inviting various Canadian stakeholders to discuss opinions and science regarding environmental issues of metals.

The workshop was organized and run by the Canadian Network of Toxicology Centres (CNTC). Attendees represented the public, private and academic sectors and included chemists, biologists, earth scientists and policy makers. Workshop participants identified some issues where consensus existed, but most of the workshop was spent listing areas of disagreement, discussing the existing science and, most importantly, identifying knowledge gaps, which, if filled, would provide extensive help in resolving the disagreements.[8]

This workshop nurtured a cooperative atmosphere and all participants voiced their concern over the lack of data and their commitment to obtaining sound data. The participants concluded that it would be beneficial to pool money and human talents together in defining a research program that could fill the highest priority knowledge gaps. A steering committee was formed to refine a detailed proposal and to negotiate funding. This committee consisted of representatives from Natural Resources Canada (Geological Survey of Canada (GSC)), Environment Canada (Atmospheric Environmental Services and the National Water Research Institute), Fisheries and Oceans, several university environmental scientists and MAC.

THE METALS IN THE ENVIRONMENT (MITE) RESEARCH NETWORK

Dealing with complex chemical and physical issues was a daunting task. This task was made more difficult by the number of stakeholders represented. Nevertheless, the committee's work was successful primarily because each member cared deeply about the outcome and was convinced that an integrated approach was far more valuable than independent activities.

It was clear that existing programs funded by Environment Canada and Natural Resources Canada were to be continued and, in some cases, extended. It was also clear that, while industry might join to support these government programs, it was essential to involve the expertise located at academic institutions across Canada. The committee also recognized the need to integrate the effort with similar activities being carried out around the world. The organization scheme was developed so that independent departments would remain primarily independent, but would be linked together in terms of research priority setting, problem formulations and sharing information.

A new NSERC Network research program was identified and preliminary meetings with NSERC officials pointed to a match between the contemplated research on metals in the environment and the mandate of the NSERC program. A criterion for NSERC was that its funding could be provided only to university research with a significant training component. It therefore became necessary to define a separate university network funding and reporting scenario. This university network will be joined to, but funded separately from, the other research programs conducted under the larger MITE umbrella, as depicted in Figure 1.

AREAS OF FOCUS

The questions concerning metals in the environment logically fall into three areas:

1. Where do metals come from? < SOURCES

2. How do metal species become altered and move between environmental compartments? < PROCESSES

3. What are the ultimate fates of "metals" and what specific harmful effects do they cause? < EFFECTS

Sources: There exists some understanding of total amounts of metals released from mining and production facilities in Canada, [9] but we have little knowledge of the species emitted to the air or to water. Even less is known about characteristics such as particle size distribution, which is expected to play an important role in chemical reaction kinetics for particles in the atmosphere prior to their wet or dry deposition. And there remains much more to be learned about long-range atmospheric transport of metals, both volatile mercury and relatively non-volatile metals.

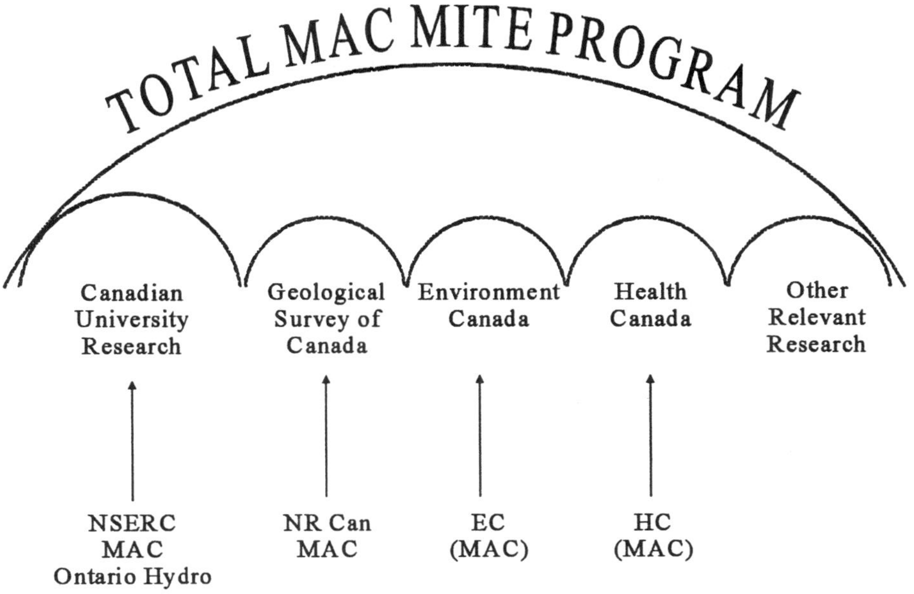

Figure 1: Total MITE Research Network and Funding Partners

Releases from natural sources are largely crude estimates at best. Releases to the soil and aquatic environments as a result of chemical weathering of natural materials will be studied in order to determine the rates of release, and the factors controlling those rates, for a range of trace elements. The knowledge gained in estimating chemical weathering rates and the fluxes from geological compartments to the aqueous environment will assist in improving natural release estimates of metals for trace metal balances in Canada. These balances form the focus of discussions on the relative importance of natural and anthropogenic inputs.

Processes: Between the sources of metals and their ultimate fates in environmental compartments lie a large number of processes by which various metal-containing species undergo reactions and are transported within and between compartments. Projects will be carried out to determine how metals from all sources are cycled by terrestrial vegetation and what factors, both biotic and abiotic, control metal partitioning among woody and herbaceous material. Metal accumulation in trees, for example, will be examined as a function of time by means of growth ring analysis and compared with historic soil metal levels. Phytochelatins in tree foliage will be evaluated as possible biomarkers of plant responses to metals.

Water movement is the driving force behind the transport of metals within the soil column; it follows that the partitioning of metals between the pore water and the solid phase is of crucial importance. A key component in soil-water equilibria is natural organic matter, both as a sorbent and as a metal-complexing/metal-reducing ligand in the mobile phase. Research is proposed to refine current estimates of partitioning coefficients and to incorporate these constants into models designed to predict the behaviour of metals in the terrestrial environment and their movement into the aquatic system.

Once metals leave the terrestrial environment, they tend to accumulate in aquatic sediments. Metal enrichment in the surface layers of the sediments of remote lakes is often interpreted as evidence for long-range atmospheric transport and deposition of metals.

However, in the current debate on the pervasiveness of metal "contamination" of the Earth's surficial environment, there is a marked lack of consensus regarding the influence of diagenesis on metal stratigraphy in lake sediment cores. Evaluation of the importance of post-depositional mobility of metals in lake sediments is clearly a priority and research will be carried out on this topic.

Effects: The aquatic environment and aquatic sediments are important sinks for metals introduced into the surficial environment. Quantitative linkage will be sought in the MITE program between:

- speciation/partitioning in the exposure media (e.g. freshwater);
- metal accumulation by exposed organisms;
- metal speciation within exposed organisms, focusing on questions of essentiality and metabolic detoxification;
- metal-induced effects at the individual, as well as population level, with emphasis on field studies related to areas of high input due either to natural or anthropogenic activity.

ORGANIZATION OF MITE

The network organizational structure is shown in Figure 2. The stakeholders selected Dr. Leonard Ritter as Program Director with Secretariat services provided by the Canadian Network of Toxicology Centres, of which he serves as Executive Director. The Program Director will report to the MITE Board of Directors, which will consist of representatives from government, industry and non-governmental organizations. The Board will be advised by a group of peers having no direct relationship with any of the research being conducted under MITE. This group will assist in evaluating priorities and research effectiveness in providing needed information.

The MITE program will be administered via three areas of focus and each area has co-leaders, one from a Canadian university and the other from a Canadian government department. The co-leaders of each area co-ordinate the preparation of proposals and monitor progress of those projects accepted for funding.

One of the prime objectives of having senior stakeholders represented on the Board is that MITE research must be carried out in a way that is acceptable to all stakeholders and that will generate results needed for drafting and implementing both government policies and industry risk management strategies.

FUNDING

The MITE university network is applying for funding to the Natural Science and Engineering Research Council (NSERC) and industry. The mining companies of MAC are contributing a total of $1.25 million over the 5-year program. Ontario Hydro is contributing $0.5 M to the program. The NSERC proposal amounts to about $3.2 M. This total amount of about $5 M will go to support only Canadian university projects within the program's scope. Other programs conducted under Natural Resources, Environment or Health departments, as shown in Figure 1, will be funded separately. Industry has also committed to joining in the funding responsibilities for those programs.

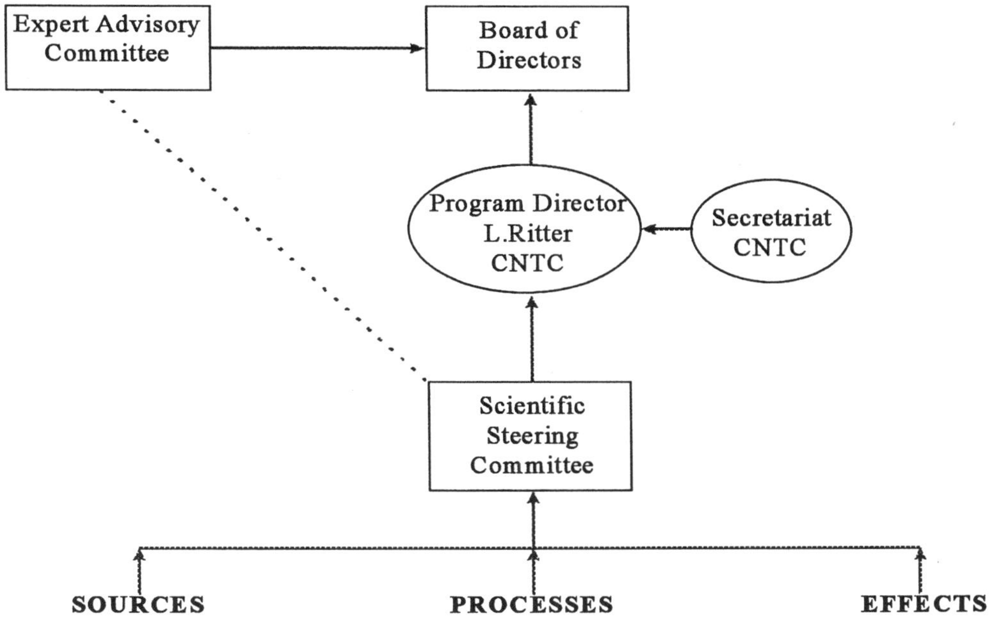

Figure 2: NSERC-MITE Organization

TRAINING

Both industry and government are in need of qualified personnel to help direct resources to solve problems that represent the highest risk for society. Inorganic substances have received only modest attention in environmental science education because of the historical emphasis on organic pollutants. Over the next several decades it is anticipated this emphasis will shift to metals and their compounds. Unfortunately, many regulators, having grown up in the synthetic organic paradigm, believe that all other substances can be simply shoe-horned into existing criteria and test protocols developed for organics. If this were to occur, much wasted effort would be spent because inorganic substances are substantially different than synthetic organics and people in responsible positions must be technically trained to appreciate such differences and to understand the scope of challenges that lie ahead as society tries to develop procedures to ensure the safe use of metals.

Training of people within MITE is particularly beneficial because of the chance to work on challenges in the real world, with field studies being an integral part of the research, within a group having a wide variety of backgrounds and with interests in multi-disciplinary approaches to problem-solving.

SUMMARY

Metals will continue to be used by society because of their specific properties. Hazards associated with metals and metal compounds must be recognized and steps taken to minimize the risk of harming individuals, human and animal populations, and ecosystems.

Metal and metal compounds are distinctly different than synthetic organic pollutants because metals occur naturally, many are essential to life, and the processes controlling their speciation, ultimate fate and toxicity are complex. MITE is a 5-year research program being jointly funded and managed by government and industry. It seeks to provide information urgently needed for carrying out sound risk assessments and developing management strategies for metals in Canada and around the world. Such data are also needed by governments as they seek to respond in a scientifically sound fashion to environmental concerns with finite human and capital resources.

The MITE program is an attempt to modify the adversarial positions sometimes taken by government and industry and to promote between these parties cooperative fact-finding and consultative decision-making so that the common goal of responsible and sustainable use of metals can be achieved.

ACKNOWLEDGEMENTS

The author wishes to acknowledge the work of MAC and CNTC staff in attending to MITE through its inception and to the Steering Committee members, who spent many long hours debating priorities and program management issues. The authors of the MITE NSERC proposal are thanked: Peter Campbell, Beverley Hale, Adrienne Laroque, Robert Garrett, Ron Pierce and Rod Allan.

REFERENCES

1. OECD Workshop on Aquatic Toxicity. Testing of Sparingly Soluble Metals, Inorganic Metal Compounds and Minerals. Ottawa, Canada. Sept. 5-8, 1995.

2. Biodegradation/Persistence and Bioaccumulation/Biomagnification of Metals and Metal Compounds. A Technical Workshop held in Brussels, Belgium, December 11-13, 1995, under the auspices of the Canada/EU Metals and Minerals Working Group.

3. International Workshop on Risk Assessment of Metals and Inorganic Compounds. Report published by the International Council of Metals and the Environment. Angers, France. November 13-15, 1996.

4. V. M. Shorrochs and B. J. Alloway. Copper in Plant, Animal and Human Nutrition. Copper Development Association. TN35, 1985.

5. J. O. Nriagu and J. M. Pacyna. "Quantitative assessment of worldwide contamination of air, water and soils by trace metals". Nature, 333, p.134-9, 1988.

6. P. E. Rasmussen. "Trace Metals in the Environment: A Geological Perspective". Geological Survey of Canada Bulletin 429, 1996.

7. I. Thornton. Metals in the Global Environment: Facts and Misconceptions. International Council on Metals and the Environment, Ottawa, 1995.

8. Final Report: Metals in the Environment Workshop, October 21-22, 1996, issued by the Canadian Network of Toxicology Centres, Guelph, 1996.

9. Summary Final Report, 1994 National Pollutant Release Inventory, issued by Environment Canada, 1996.

The International Lead Management Center:
An Industry Commitment To Cooperation

C.J. Boreiko
Executive Director
International Lead Management Center
2525 Meridian Parkway, suite. 100, P.O. Box 14189
Research Triangle Park, NC U.S.A. 27709-4189

ABSTRACT

Traditional patterns of natural resource utilization are now being evaluated in accordance with new paradigms which seek to ensure the long-term compatibility of global industrial activity with preservation of human health and the environment. As a consequence, the lead industry has experienced intense scrutiny from the international community and numerous regulatory options have been considered which would govern international industry practice. In response to concern expressed by the OECD, the international lead industry engaged in a cooperative dialog which resulted in the adoption of non-regulatory instruments as the preferred course to achieve international lead risk reduction goals. This paper summarizes some of the nonregulatory outcomes implemented to address the concerns of the international community and, as such, may serve as a useful guide other nonferrous metal sectors.

Proceedings of
Metals and the Environment: An International Symposium
Edited by Jim Skeaff
The Metallurgical Society of CIM
100th CIM Annual General Meeting, 1998

INTRODUCTION

The international community is increasingly focused upon environmental issues, with particular attention being directed towards patterns of natural resource utilization which are compatible with sustainable development. The nonferrous metals industry is one industry sector currently under scrutiny, with questions being asked regarding the environmental impacts associated with both primary and secondary metal production. Measures of industry performance are no longer based solely upon compliance with regional regulatory frameworks. Rather, the production of base metals traded in international commerce is being evaluated from a perspective which implicitly imparts accountability for the aggregate environmental performance of the global industry. End use applications of metals are under similar scrutiny, with international agencies seeking to ensure that the use and/or end-of-life disposition of metal-containing products does not pose undue risk to human health or the environment. Once again, the performance measures being applied seek to ascertain whether adverse ramifications are associated with metal utilization by the international community in a diverse range of products. Simply put, traditional industry practices and patterns of natural resource utilization are being evaluated in accordance with new paradigms which seek to ensure the long-term compatibility of global industrial activity with harmonized international goals for the preservation of human health and the environment.

The rapid development of international instruments seeking to address environmental issues has ramifications for virtually all non-ferrous metals. Recent debates regarding lead provide examples of how the concerns of the international community can potentially impact upon basic patterns of metal production and usage. Ongoing discussions concerning the Basel Convention will likely affect materials flow for numerous scrap metal categories in international commerce. Concerns over Long Range Transport of Air Pollutants within the United Nations Economic Commission for Europe may affect performance standards for metal production facilities in numerous countries, as well as regulate lead utilization in consumer products which enter the municipal solid waste stream for possible incineration. Efforts to harmonize classification and labeling criteria by agencies such as the Organization for Economic Cooperation and Development (OECD) may similarly impact upon lead usage in consumer products and dictate guidelines for the transboundary movement of bulk materials. Specific examination of lead by the OECD Risk Reduction Program resulted in consideration being given to implementation of broad-based prescriptive measure regulating lead production and use throughout OECD member countries.

There are those who prefer to regard the environmental issues surrounding lead as unique, and the international attention focused upon it as an aberration. This viewpoint overlooks the parallel efforts which have been impacting upon the production and use of cadmium and mercury, and fails to recognize that the international community has now developed instruments for achieving concerted action to counter perceived risks from virtually any material. Moreover, virtually all metals in commerce are perceived to pose a risk to human health or the environment by one or more national governments influential in the international community. The experience of the lead industry is thus a foreshadowing of the scrutiny likely to be experienced by other metal industry sectors. Examination of recent events specific to lead is instructive, both with respect to the new expectations being placed upon the private sector by the international community and the opportunities which are created when industry engages itself in a meaningful discussion of the issues.

THE OECD RISK REDUCTION PROGRAM

In 1990 the member states of the OECD undertook a Pilot Program in Risk Reduction, evaluating the need for concerted international action to address the risks posed by five substances. Lead emerged to be the most commercially important and intensely debated substance under evaluation in the Risk Reduction Program - stimulating a six year exchange of information and viewpoints between industry and the OECD membership. Industry's willingness to engage in dialog, and the willingness of the OECD membership to consider the input of industry, resulted in a outcome to the OECD Lead Risk Reduction Program somewhat different from the broad prescriptive measures initially proposed to harmonize international regulations concerning lead industry practice and lead product applications.

On 20 February, 1996 the Environment Ministers of the OECD issued a Declaration on Lead Risk Reduction seeking to develop and strengthen national and cooperative efforts considered necessary to reduce risks from exposure to lead. They proposed a process which took into account national priorities, policies, programs and achievements, recognizing that implementation may take the form of either voluntary or regulatory actions, or both. The Declaration recognized the differing needs and circumstances of member and non-member countries and called for flexible national risk reduction strategies and time frames. The highest priority was suggested for actions which addressed the risk of exposure from food and beverages, water, air, and occupational exposure. The Declaration also initiated a process which would continue to review lead levels in the environment and exposure to lead for sensitive populations such as children.

The Declaration further proposed the extension of cooperative efforts to share information about exposures of concern, risk reduction options and environmentally sound and economically viable technologies useful to reduce risk from exposure to lead. In this context the lead industry was invited to share their experience in the sound management and prudent use of products containing lead, including the use of alternative materials in specific product applications. The lead producing and using industries were encouraged to make the best use of their expertise in the management of lead risk reduction and asked to make this expertise available to OECD member and non-member countries.

The international lead industry recognized that exposure to lead can result in risk to human health and the environment. The nature and extent of these risks, and the measures which had been adopted by OECD countries to address them, had been comprehensively reviewed by the OECD in the Risk Reduction Monograph issued in May of 1993 and the OECD Toronto Workshop on lead products and uses of September 1994. Both the Monograph and the Toronto Workshop demonstrated the significance of different exposure sources and how they vary from one country to another. The effective management of risk thus required programs tailored to national or regional circumstances, needs and priorities. Although important lessons could be learned from the experience of others, different countries implementing lead risk reduction initiatives would achieve optimal results if the programs adopted took into account their particular environmental factors and economic, social and cultural circumstances.

To address the needs of both OECD and non OECD countries, the international lead industry proposed to undertake a voluntary risk reduction action program consistent with the recommendations of the United Nations Council for Economic Development, Agenda 21, for action by governments and industry to enhance the sound management of chemicals. The industry's proposal would not duplicate, but rather complement and support the activities and needs of international agencies and countries. Flexibility would be provided by developing and conducting initiatives on a project by project, country by country basis. Execution of this voluntary industry risk reduction project was placed under the auspices of the newly formed International Lead Management Center (ILMC).

THE INTERNATIONAL LEAD MANAGEMENT CENTER

The International Lead Management Center was founded in July 1996 with headquarters in Research Triangle Park, North Carolina, USA. The Center was established by the following major mining and lead producing corporations.

Aberfoyle-Australia :a mining and exploration company producing zinc, lead, and silver concentrates.

ASARCO-USA: one of the world's leading producers of nonferrous metals, principally copper, silver, lead and zinc.

Boliden-Sweden: a thoroughly integrated smelting and refining company for pure ores as well as complex and contaminated raw materials.

Cominco-Canada: one of the world's largest producers and marketers of zinc concentrates and lead metal.

Doe Run-USA: "the life cycle company", is North America's largest integrated lead mining, milling, smelting and recycling company

Met-Mex Peñoles-Mexico: one of Mexico's main natural resources and industrial groups.

MIM Holdings-Australia: a major international metals and energy company which mines and produces gold, copper, silver, lead and zinc.

Noranda-Canada: one of the world's largest producers of zinc and nickel, and major producers of copper, lead, gold, silver and potash.

Pasminco-Australia: one of the world's largest vertically integrated base metal companies producing lead and zinc concentrates, lead, zinc and silver metals as well as the various alloys and by-products.

Union Miniere-Belgium: a world leader in the nonferrous metals sector, UM produces and sells more than 20 nonferrous metals for basic and advanced technology industries, the construction sector and research.

The Terms of Reference and activities of the ILMC are under the auspices of an Executive Board composed of representatives from the member companies. In an effort to gain additional guidance and direction on priorities, project feasibility and action programs, ILMC also assembled an independent Policy Advisory Group (PAG) comprised of independent experts with international experience in lead mining, smelting, refining and recycling, occupational and public health, risk reduction programs and technology transfer. Most PAG members are, or have been, associated with national regulatory agencies or intergovernmental agencies, but provide guidance in a personal capacity. This PAG serves to provide independent assessment and direction for the ILMC and helps ensure that ILMC activities are consistent with the needs of the international community and the requirements of the OECD Declaration.

THE ILMC PROGRAM OF WORK

One of the Center's primary activities is the coordination of industry expertise. The membership of ILMC possesses expertise in a variety of risk reduction areas - making the expertise of the industry available as appropriate to address technical inquiries or to actively participate in risk reduction efforts is one of the important coordinating functions of the ILMC. Calling upon this expertise, the ILMC has been initiating a program of work which recognizes the specific goals and objectives of the Ministerial Declaration on Lead. The following goals were noted to be of specific interest to the OECD member countries:

- a progressive phase-down in the use of lead in gasoline except where needed for essential or specialized use for which there is no practical or viable alternative.
- eliminate the exposure of children to lead resulting from products intended for use by children (e.g. toys, cribs, crayons).
- eliminate exposure to lead from food packaging and cans through either substitution or elimination and wherever this was impractical to use functional barriers to prevent lead migration.
- restrict exposure of lead from leaching from ceramicware and glass crystal used for food and beverages by either a change in production methods or process controls.
- restrict the use of lead shot in wetlands and promote the use of alternatives to lead sinkers in shallow waters.
- reduce lead levels in drinking water through appropriate treatment measures and the use of materials in distribution systems which do not release lead into water.
- reduce levels of occupational exposure by limiting atmospheric emissions from major point sources, and to establish strategies, including public information programs, to reduce significantly exposures arising from the historical use of lead containing materials in buildings.
- phase-down the use of lead in paint and rust proofing agents except in cases of essential or specialized uses for which there is not practical alternative.

ILMC was founded on the assumption that industry has a valued and positive role to play in the achievement of these goals. Towards this end, a series of activities were initiated to address both country-specific risk reduction needs and the general information requirements of the international community. Specific ILMC initiatives include the following.

Pilot Programs

ILMC is undertaking a series of specifically designed Risk Management Processes, or Pilot Programs, to demonstrate country-wide approaches to risk reduction. Pilot Programs address the distinct lead exposure issues unique to each country's cultural, technical, geographical and socioeconomic circumstances. ILMC Pilot Programs are essentially joint cooperative undertakings in which ILMC, usually upon invitation from a national government, assists in the identification, planning and implementation of lead risk reduction measures appropriate for that country. Pilot programs are constructed as multi-stakeholder consultative efforts involving government, industry and representatives of environmental and/or public health sectors. These programs entail a risk management process which is underwritten by a commitment of all parties to reduce lead exposure.

When ILMC receives a request for assistance from a government agency or industry sector wishing to initiate a lead risk reduction program, a team of ILMC experts is assigned to collect, collate and analyze all the available data pertaining to the lead exposure issue of concern. This team also establishes an information and technology exchange network with the international community. All investigative data, research and analysis and program documentation, including any training modules are, however, retained by the government agency and those industries responsible for implementing Pilot Programs.

Though a step-wise process, Pilot Programs seek to define appropriate targets for industry performance and are structured so as to produce tangible benefits to individual countries within a time frame of several years. However, the process pursued is of equal importance in that it seeks to evaluate and document the effectiveness of different strategic approaches to risk reduction. Through the implementation of Pilot Programs, the ILMC seeks to establish guidelines for the construction of voluntary agreements to serve as models for industry and governments in other countries. Towards this end, careful documentation has been made of the steps taken in each country and it is anticipated that capacity building "case studies" will be published upon the completion of each effort.

ILMC Pilot Programs are ongoing in Mexico and the Philippines, each exploring different strategic approaches to risk reduction. Opportunities for the implementation of Pilot Programs in an additional five countries are under evaluation, with the expectation that at least two additional initiatives will be underway in 1998. Candidate Pilot Programs are being evaluated with attention given to both geographic diversity and the nature of the risk reduction challenges presented. The status of ongoing programs is as follows:

Mexico

Activities in Mexico have focused upon the creation of sector-wide voluntary initiatives involving the majority of lead producing and using industries in Mexico. The structuring of this effort has progressed through a series of stages, each obtaining broader commitment to the risk reduction process from an ever increasing circle of stakeholders. Preliminary consultations for a multi-stakeholder Pilot Program in Mexico were initiated in late 1996 and meetings held involving representatives of ILMC, Mexico's National Institute for the Environment and the Mexican Chamber of Mines. These preliminary consultations outlined a multi-stakeholder consultative process which would be pursued in a Pilot Program.

Following these initial consultations, a "stakeholders meeting" was convened in Mexico City and included representatives from ILMC and industry sectors concerned with the production and use of lead in Mexico. This stakeholders meeting produced agreement to move forward with efforts to develop a voluntary lead risk reduction Pilot Program in conjunction with ILMC, the government of Mexico, the Mexican Chamber of Mines and the private lead producer sector.

A two-day workshop was then convened on April 29-30 in Monterrey, Mexico to discuss and consider lead risk reduction needs. The two-day workshop consisted of morning plenary presentations which covered lead production and use in Mexico, and internationally recognized exposure control regimes. The morning plenary presentations were followed by "breakout groups" to discuss the issues presented during the morning sessions and consider actions plans for risk reduction programs. The principal topics of discussion at the workshop included:

1) Reducing environmental risks from airborne process emissions;
2) Reducing risks from lead-bearing liquid effluent and solid waste residues;
3) Reducing the risks of lead exposure in the workplace;
4) Biological monitoring in the workplace.

ILMC has since formalized agreements with the government establishing the process by which desired improvements in environmental and occupational performance will be pursued. Risk reduction is being pursued on a sector by sector basis, with demonstration risk reduction projects being implemented at a range of industrial facilities. Networking strategies are being implemented to communicate those technologies which prove to be effective in each sector and to assist the stakeholders in identifying and documenting continuous improvement in environmental performance.

Philippines

ILMC activity in the Philippines was undertaken as an exercise in cooperation with intergovernmental agencies to address a single sector-specific issue of concern to a large geographic region. ILMC has been working with the United Nations Conference on Trade and Development (UNCTD) regarding the need to evaluate and improve the environmental performance of secondary lead recycling facilities in the Philippines. The Philippines is one of a number of countries in the Pacific region experiencing active growth in demand for lead. Current consumption demands are mostly met through secondary lead production. Moreover, current rates of growth suggest the need for an environmentally sound recycling industry and battery collection infrastructure.

ILMC has been working with UNCTAD in the structuring of multi-stakeholder discussion panels to evaluate, and make recommendations regarding, occupational and environmental aspects of secondary lead production in this country. The ultimate goals of this effort are the identification and evaluation of country specific options for ensuring sound environmental performance in the recycling industry. UNCTAD has received commitments from Philippine government agencies and the major recycling companies to this process. Academia and environmental groups are among the stakeholders being included in the upcoming consultations.

ILMC is presently assisting in the preparation of a series of background papers which will facilitate the efforts of the discussion panels to evaluate the need for improvement in occupational and environmental performance and the costs and benefits of improved performance. ILMC expertise is also being made available as needed for assessments of the environmental performance of existing facilities and will be a participant at the discussion panel meetings. The first of these meetings is currently being scheduled for mid 1998.

ILMC Information Resources and Training Materials

In order to serve the general needs of the international community, and to assist in the implementation of pilot programs, ILMC has been developing a variety of information resources regarding industry practice and different product applications. Due to the anticipated emphasis of ILMC Pilot Programs upon developing countries, highest priority has been placed upon issues of concern to non-OECD countries.

The ILMC membership has extensive experience in implementing effective lead risk reduction strategies and is in a position to offer a comprehensive range of training and tutorial programs. For those managers charged with the responsibility for reducing lead exposure ILMC has prepared, or is preparing, development courses, seminars and workshops covering such topics as the management of change, controlling and changing attitudes, problem solving, data analysis, and the design of effective facility management solutions. For occupational health and environmental management practitioners, specific modules are available to improve awareness and examine successful case histories in reducing lead exposure in the workplace. ILMC has also been generating information specific to individual product application.

Clearinghouse for Lead in Gasoline Phase-out Information

Many countries are attempting to establish an adequate public health policy to reduce the use of lead-containing additives in automotive fuels. ILMC participated in multi-stakeholder consultations convened by UNEP-Industry and Environment and OECD. As a consequence of these consultations, ILMC is establishing a clearing-house on information pertaining to lead in gasoline. This clearinghouse function entails the preparation of an annotated bibliography as a guide to countries formulating public health policy on leaded fuels and the distribution of informational materials. In addition, the clearing-house will contain information packages such as databases on engine technology capable of using unleaded fuels, existing oil refinery capacity and technology, and information on the composition of international fuel stocks. Finally, ILMC is working with UNEP-IE and OECD to prepare a series of "consensus issues papers" on topics such as the impact of older vehicles upon phase-out strategies and the relative merits of implementing different phase-out policies. The clearing-house will be maintained on a dedicated ILMC web site, with linkages to the site being provided by OECD, UNEP and others.

Lead Product Information

ILMC is working with end-user industries and associations in the preparation of technical guides and manuals concerned with the safe manufacture and use of lead in specific product applications. Ongoing efforts are focusing upon lead cystalware, ceramicware and lead shot. Each product application is being addressed in full consultation with the end-user industry sectors involved.

ILMC is also working with the United Nations Industrial Development Organization (UNIDO) and OECD to address issues associated with other lead product applications. Information transfer tools tailored to the needs of developing countries are under development in a demonstration project for lead soldered food cans.

Other Communication Mechanisms

Information systems and communication media are key to the success of the ILMC. An ILMC brochure is available and has seen wide distribution at major industry, health and environmental meetings and conferences. The brochure can be customized to accommodate specific requests and needs by the insertion of specially prepared information "fact sheets" which provide brief summaries on risk management and lead product issues. The brochure serves as a vehicle for both enhancing awareness of the Center and providing preliminary information transfer on specific lead risk reduction concerns.

A quarterly newsletter, "NewsCasting" was initiated in June 1997, with broad distribution to a mailing list of over 500 individuals and groups interested in reducing the risk of lead exposure and the activities of the Center. The newsletter provides recipients with regular updates on the risk management activities of ILMC and the lead industry. ILMC has also initiated a web site (www.ILMC.org) to promote awareness of the services of the center, information transfer and inquiries.

SUMMARY

ILMC has embarked upon a program of work attentive to the range of lead risk reduction issues identified in the OECD Ministerial Declaration and is making a significant contribution to lead risk reduction in many areas of the world. Ongoing ILMC activities have significantly enhanced the capabilities of the Center and the services offered will grow with time. In every project undertaken, ILMC works in partnership with governments and industry to develop risk reduction strategies and tools to ensure the success of international lead risk reduction efforts. The Center, through its actions, has sought to provide affirmation of the international lead industry's commitment to cooperation and to respond to the needs of individual countries and industry sectors. As environmental issues rise in the international theater, new challenges will be posed regarding issues of sustainable development and the non-ferrous metals industry. The ILMC stands as one indicator of the lead industry's commitment to be a contributor, as opposed to an adversary, to the international processes which will shape the policies that will govern future natural resource utilization and reuse.

Resolving metals issues – the way ahead

Mining Nature

Monte Hummel
President and CEO
World Wildlife Fund Canada
90 Eglinton St. E., suite 504
Toronto, Canada M4P 2Z7

ABSTRACT

This presentation will explore the real versus imagined areas of disagreement between the mining industry and wilderness advocates, using specific examples.

While there have been areas of disagreement in principle, initiatives such as the Whitehorse Mining Initiative reflect a willingness by both parties to resolve some of these. And while there certainly have been specific examples of head-on conflicts, these "Mexican stand-offs" have been wrongly used (by both parties) to characterize the entire relationship between miners and conservationists. There are equally many examples of wilderness advocates and mining interests working together to accommodate each other, and it's time these examples were spotlighted and laid squarely before those on both sides who are intent on playing off old stereotypes.

Proceedings of
Metals and the Environment: An International Symposium
Edited by Jim Skeaff
The Metallurgical Society of CIM
100th CIM Annual General Meeting, 1998

Sustainable development and mining: A practical concept that yields a positive return on investment

B.W. Feltmate
President
Sustainable Systems Associates Limited, Suite One
1803 Storrington Street, Pickering, Ontario, Canada L1V 2X2
tel: 905-420-9956, fax: 905-420-1426, feltmate@idirect.com

ABSTRACT

Sustainable development is inclusive of stewardship practices which address the three interdependent components of environment, economy and society. Within the mining sector sustainable development can be defined and operationalized in practical terms. For leading-edge mining companies, 25-40 measures per each category of environment, economy and society are common.

The Return on Investment (ROI) conveyed to companies that are practitioners of sustainable development can be assessed anecdotally and quantitatively. The quantitative demonstration of ROI (in this paper referenced against the TSE 300) is presented as an emerging driver to sustainable development best practices for companies.

Proceedings of
Metals and the Environment: An International Symposium
Edited by Jim Skeaff
The Metallurgical Society of CIM
100th CIM Annual General Meeting, 1998

INTRODUCTION

Following the United Nations Conference on Environment and Development in 1992, over 170 countries agreed on the need for businesses to incorporate sustainable development into their operations. Since this time, many companies from all industry sectors, including mining, have committed themselves to sustainable development. Why?

In brief, corporations operate with the consent of society. Within the mining sector, prior to the 1990s, substantive environmental programs and practices served companies well in meeting society's expectations. However, society's expectations have changed, and a focus on sustainable is now expected. Thus, society's expectations are evolving, and so too is, and must, the mining industry.

The term sustainable development is the now familiar "development that meets the needs of the present without compromising the ability of future generations to meet their own needs", as expressed in the world acclaimed publication "Our Common Future", otherwise referred to as the Brundtland Commission Report (WCED 1987). Sustainable development consists of three interdependent components -- environment, economy and society -- and is therefore, by definition, more encompassing than its environmental antecedent (SMAC 1998).

Although many companies have a general sense of sustainable development, when the time for implementation is actually "at hand", two questions inevitably arise:

- how can sustainable development be operationalized in a manner that is practical, meaningful and cost-effective, vis-à-vis initiatives within the spheres of environment, economy and society?
- what is the Return on Investment (ROI) for being a practitioner of sustainable development? How will/does sustainable development contribute to the creation of shareholder value?

The following discussion will address these questions in the order in which they appear. It is necessary to first understand sustainable development from an operational perspective before calculating the ROI.

DEFINING AND OPERATIONALIZING SUSTAINABLE DEVELOPMENT

The first step in operationalizing sustainable development is to define the concept in a manner applicable to a particular company. The definition should reflect the "spirit" of Brundtland. For example, relative to mining, sustainable development might be defined as:

> "to design, construct, operate, decommission and monitor mines in a manner that meets the environmental, economic and social needs of present and future generations."

Some companies may wish to identify particular stakeholders by replacing the word "present" with "shareholders, employees, customers, and the residents of local communities".

There is no absolute definition of sustainable development. What is appropriate for one company may not be for another. Despite the definition adopted by a company, it should reflect a commitment to current and future generations, and the interdependent components of environment, economy and society.

Most companies reach consensus on defining sustainable development by means of a round table discussion. Internal participants generally include the company's environmental, operations and communications personnel, and external groups can include NGOs, aboriginal groups, consultants, and national and international professional associations and governmental representatives.

Once a corporation has defined sustainable development, the more perplexing challenge is "what do you do next?"

Many leading-edge corporations proceed by establishing a declaration of six-to-ten principles of sustainable development, which reference the three interdependent components of environment, economy and society. Three examples of such principles are illustrated in Table 1.

TABLE 1: Examples of Sustainable Development Principles Which Address Issues of Environment, Economy and Society[a]

Category	Principle
Environment	• We will minimize the physical, chemical and biological effects of our activities on the environment.
Economy	• We will maximize the creation of shareholder value and ensure the fair distribution of financial benefits to all stakeholders.
Society	• We will facilitate constructive dialogue with stakeholders in order to anticipate and address issues of regional and global sustainability that are relevant to our business.

[a] these principles are currently in use within multinational, Canadian-based mining companies.

When formulating principles of sustainable development, it is important to bear in mind that they will only be as effective as a company's ability to establish practical, meaningful and cost-effective practices.

The definition of sustainable development, when combined with six-to-ten principles of sustainable development, can constitute a company's Sustainable Development Policy. This statement is usually a one-page, stand-alone document (e.g., see the Sustainable Development Policy of Falconbridge Ltd. and Placer Dome Inc.), that presents a concise encapsulation of the company's commitment to environment, economy and society. Within a company the policy would generally be placed in strategic locations frequented by high numbers of employees, and externally it would be distributed to a wide range of stakeholder groups.

Establishing measures that address environmental principles of sustainable development is generally not difficult for corporations, particularly in the mining sector. There is a rich history of expertise in this area. The greater challenge resides in identifying social and economic measures. To illustrate, Tables 2 and 3 present examples of "typical" social and economic measures, respectively.

When reviewing the measures identified in Tables 2 and 3, it should be evident that any mining company that has conducted itself as a "good corporate citizen" may already be addressing issues of social and economic relevance, in additional to having implemented a well established complement of environmental measures. Thus, sustainable development should not be viewed as a completely novel pursuit for companies that already conduct themselves responsibly. In such cases, sustainable development should be viewed as a "driver to

TABLE 2: Examples of Social Measures Applicable to Mining Operations[a]

General Description	Measure
Community Outreach	• Meetings with residents of local communities, Chamber of Commerce, NGOs, etc., to review concerns from environmental, economic and social perspectives. An example of an issue for discussion could be the affect of mining on local tourism and recreation • record No. of meetings with residents/year, specific actions resulting from meetings, No. site visits by interest groups/ year, total No. of visitors/site/year, etc.
Historical and Archeological Sites	• efforts to preserve and restore historical sites • efforts to preserve and restore archaeological sites • record specific sites preserved and restored • note efforts to support archeological programs as per universities, government agencies, etc.
Noise Pollution	• minimize noise pollution • note efforts to minimize pollution (e.g., cowling around fans)
Light Pollution	• minimize light pollution • note efforts to minimize pollution (e.g., reflectors on lights)
Odour Pollution	• minimize odour pollution • note efforts to minimize pollution (e.g., no releases of H_2S)
Aesthetic Values	• efforts to maintain aesthetic quality of resources (e.g., landscaping tailings and waste rock areas, planting trees to obscure "unpleasant" view of operation, etc.) • note efforts to maintain aesthetics
Employee SD Practices	• transfer sustainable development practices to communities (e.g., provide recycling expertise to communities) • note efforts to promote transfer of best practices

[a] 25-40 social measures characterize a typical resource-based company practising sustainable development (SMAC 1998)

behaviour" that can build on current momentum by pushing companies further along the environmental, social and economic continuum.

RETURN ON INVESTMENT (ROI)

Anecdotal Evaluations of ROI: The ROI for sustainable development can be expressed anecdotally, as per Table 4. Relative to mining, the first issue -- Better Relations with Residents of Local Communities -- is a primary driver to sustainable development best practices. For example, anyone familiar with the mining industry can list cases where community perceptions of poor corporate performance have slowed the opening of new mines, and in some cases may

TABLE 3: Examples of Economic Measures Applicable to Mining Operations[a]

General Description	Measure
Local Hiring Policy	• hire employees locally where feasible • record No. of employees hired locally
Skills Training	• provide special skills training for employees • note the development of transferable skills training
Local Procurement	• purchase goods locally where feasible • calculate total monetary figure pertaining to goods purchased locally
Vendor Procurement Policy	• encourage sustainable development best practices by vendors (e.g., help vendor develop EMS) • note vendor's sustainable development policy and practices
Share Environmental Best Practices/ Technology	• share environmental best practices with mining companies • note best practices and technologies shared with companies
Contribution to Science and Technology	• develop cost-effective, meaningful, practical and interpretable means of assessing environmental impacts (e.g., contributions to MEND, AETE, etc.) • note committee participation
Taxes	• calculate monetary figure pertaining to total taxes paid at national and regional levels • record monetary figure pertaining to total tax payment
Post-Operational Commitment	• engage meeting with local Chamber of Commerce, Mayor, citizen groups, etc., to help ensure post-operational economic activity • record efforts to assist community in establishing post-operative economy (e.g., developing tourism, retirement community, etc.)

[a] 25-40 economic measures characterize a typical resource-based company practising sustainable development (SMAC 1998).

have contributed to their cancellation. Thus, from a community perspective sustainable development is a vehicle to profile the environmental, economic and social commitment of an organization in a manner that will nurture public favour, which has obvious positive ramifications from the perspective of ROI. Other issues listed in Table 4 are self-explanatory vis-à-vis ROI.

TABLE 4: Reasons Why Companies Practice Sustainable Development (issues are presented in no particular order) (SMAC 1997)

Issue and Stakeholder	Explanation
Better Relations with Residents of Local Communities: Stakeholder ⇨ Local Residents	• the cost of poor public relations with residents of local communities can be substantial if it results in resistance to new project start-ups, facility expansions or ongoing developments.
Promote Industry-Wide Self-Regulation: Stakeholder ⇨ Industry/Government	• some regulatory procedures can cost industry large sums of money while providing limited environmental advantage. Regulatory programs designed by industry may be more cost effective and provide greater environmental benefit.
Enhance Due Diligence Protection: Stakeholder ⇨ Corporate Executives	• during 1990-1995 a record number of Canadian corporate executives were convicted of environmental offenses (2.5 executives/year). Total annual environmental fines in excess of $1 million are not uncommon in the resource sector.
Address Value Chain: Stakeholder ⇨ Customers, Suppliers, etc.	• customers are increasingly concerned vis-à-vis the harm that corporate practices might convey to the environment. ISO[a] and the International Chamber of Commerce have developed objective criteria to assess environmental product claims.
Address Media and Activist Pressures: Stakeholder ⇨ Media and Activists	• large organizations such as American Rivers, Greenpeace, Friends of the Earth, Nature Conservancy of Canada and the Sierra Club can affect public perceptions of corporate operations, which can be costly to industry.
Corporate Commitment to Stewardship Practices: Stakeholder ⇨ Employees	• any corporate sustainable development program must have the support of all employees, from the boardroom to the shop floor. Employees who are "committed" to a corporate cause will work more diligently to "realize a vision".
Lower Insurance Premiums: Stakeholder ⇨ Insurance Companies	• although few Canadian companies currently carry environmental insurance (i.e., many major companies are self-insured), insurance companies are including sustainable development initiatives in the underwriting process.
Lower Bank Loan Rates: Stakeholder ⇨ Banks	• most major banks employ senior environmental managers to assess the cumulative environmental risk associated with the lending of money for mortgage holdings, land acquisitions, etc. Interest rates are adjusted to reflect risk.

[a] ISO 14024 addresses the issue of product labeling.

Quantifiable Evaluations of ROI: It is now possible to quantify the sustainable development performance of corporations as per a Sustainable Development Index (SDI©: Feltmate and Schofield 1997), and examine whether companies with a high SDI rating yield superior stock price performance. Thus, the ROI for sustainable development performance can now be assessed at a level beyond the anecdotal explanations presented in Table 4. Although the method to calculate the SDI of a company is beyond the scope of this paper, a brief explanation is in order.

The data to quantify the SDI of a company is derived from several sources. For example, since approximately 1995 onward, companies from nearly all industry sectors worldwide have been publishing environmental and sustainable development reports that document performance *via* quantifiable data (e.g., SO_2 emissions/year, concentration of metals discharge from tailings/year, number of community consultation meetings/year, No. local employees hired/operation, etc.). The information presented in these reports can be complemented with interviews with VPs/Directors/Managers, questionnaires, site visits, etc., to establish a holistic profile of the sustainable development performance of a company. Based on this information, an SDI algorithm can be applied to quantify the sustainable development performance of a company.

Table 5 profiles the stock price performance of 17 companies that ranked "high" as per SDI performance (Feltmate and Schofield 1997), and they are benchmarked against the performance of the Toronto Stock Exchange 300 (TSE 300) over the equivalent time period. The 17 companies comprising the SDI are derived from the following wide range of industry sectors (i.e., the SDI composite is an analogue of the TSE 300): beverage, chemical, computer hardware, forestry, mining, petroleum, steel/manufacturing, telecommunications and utilities.

TABLE 5: Percentage Increase in Stock Price of High Ranking Sustainable Development Index (SDI) Companies[a] and the TSE 300 (over time intervals Aug. 96 vs. Aug. 95, and Aug. 97 vs. Aug. 95)

Measure	% Increase Aug. 96 vs. Aug. 95 0 → 60
SDI Companies	
TSE 300	
	% Increase Aug. 97 vs. Aug. 95 0 → 60
SDI Companies	
TSE 300	

[a] companies which have operationalized sustainable development, as per environmental, economic and social measures.

As Table 5 indicates, between Aug. 96 vs. Aug. 95, and Aug. 97 and Aug. 95, the differences in percentage increase of the share price of SDI companies relative to the percentage increase of the TSE 300, were +6.5% and +10.9% respectively. This data suggests that the sustainable development performance of a company is a good predictor of a positive, high yield ROI (relative to the TSE 300). Financial institutes and organizations are beginning to document the relationship between sustainable development/environmental performance and a positive ROI for companies and shareholders (e.g., Blumberg *et al.* 1997, Figge 1997).

The relationship between SDI and ROI is probably driven by a combined cause-and-effect relationship (driven by the issues outlined in Table 4), and a correlation with good corporate leadership. For example, companies which think "ahead of the curve" in relation to sustainable development probably also do so regarding financial management, operational management, etc., which also provide a positive ROI.

SUMMARY

Sustainable development is a tangible concept that has been operationalized in many companies, including increasing numbers in the mining sector. The number of measures that address issues of environment, economy and society varies between 80-120, depending on the size of a company and the industry sector to which it belongs. For leading-edge mining companies, 25-40 measures per each category of environment, economy and society are not uncommon.

The ROI for being a practitioner of sustainable development can be assessed both anecdotally and quantitatively. The quantitative demonstration of ROI (in this paper referenced against the TSE 300) will presumably emerge as a strong, and perhaps central, driver to behaviour for companies to be sustainable development practitioners.

In sum, sustainable development is now shown to be practically achievable and financially beneficial for corporations.

REFERENCES

BLUMBERG, J., KORSVOLD, A., and BLUM, G. 1997. Environmental Performance and Shareholder Value. World Business Council for Sustainable Development, Conches-Geneva, Switzerland. 70 p.

FELTMATE, B.W. and B. SCHOFIELD. 1997. Sustainable Development Index (SDI©) Quantification. Unpublished protocol.

FIGGE, F. 1997. Environmental Shareholder Value. WWZ-Study No. 54. Center of Economics and Business Administration (WWZ), University of Basel and Bank Sarasin & Co., Basel. 37 p.

SOCIETY OF MANAGEMENT ACCOUNTANTS OF CANADA (SMAC). 1998. Writing and Evaluating Sustainable Development and Environmental Reports. SMAC, Hamilton, Ontario. 60 p.

SOCIETY OF MANAGEMENT ACCOUNTANTS OF CANADA (SMAC). 1997. Accounting for Sustainable Development: A Business Perspective. SMAC, Hamilton, Ontario. 58 p.

WORLD COMMISSION ON ENVIRONMENT AND DEVELOPMENT. 1987. Our Common Future. Oxford University Press, New York, 400 p.

Sustainable development and mining: opportunity or threat to the industry?

Wesley Cragg
Gardiner Professor of Business Ethics, Dept. of Ethics, York University
4700 Keele St., North York, Ontario, Canada M3J 1P3

ABSTRACT

A five year study of four resource extraction projects indicates that:

- articulating a clear, honest transparent and defensible vision of the values that lie at the heart of resource extraction projects is crucial to building public confidence in resource extraction industries.
- commitment to principles of sustainable development has an important role to play in reducing conflict and building public confidence and trust in companies proposing or engaged in resource extraction activities.

These two observations provide a key to discerning the way ahead for the mining industry. They also pose significant challenges. Mining in widely seen as an unsustainable industry. Historically, the industry has nourished this perception by defining its environmental, social and economic responsibilities narrowly. These optics, however, are not insurmountable obstacles. Mining is not in principle incompatible with environmental, social and economic sustainability. Defending this view is the central task of my paper.

It would be a mistake to underestimate the challenge which the mining industry faces in convincing a sceptical public that it has a valuable contribution to make to the economies of developed, developing and under developed areas of the world. This observation will be surprising to some advocates of mining who may be tempted to dismiss it as a view based on a profound ignorance of mining as it has evolved in the second half of the twentieth century. After all, research shows that the mining and the primary metals industry is a high investment, high productivity, high income sector capable of generating important foreign earnings, or so the Canadian experience would suggest (Peter Dungan, 1997). Set against this view, however, is the history of mining, the non renewable nature of the resources mining extracts and the frequently very visible impact on the natural environment that mining has. Let us look briefly at each in turn.

Mining does not have a good track record on which to build. Historically, the industry has defined its environmental, social and economic responsibilities narrowly. Where costs could be externalized they frequently have been. The legacy is one of abandoned mines, scarred landscapes, ghost towns, air and ground water pollution and so on. Neither is this just a legacy. A search through news stories this decade will provide all the ammunition that the industry's many critics need to argue that things have not changed much in recent years. There are counter arguments, of course. The Whitehorse Mining initiative is a good illustration of this fact. But trust once undermined is not easily restored. And bad examples are easily found even if not always accurately generalized to the industry as a whole.

Proceedings of
Metals and the Environment: An International Symposium
Edited by Jim Skeaff
The Metallurgical Society of CIM
100th CIM Annual General Meeting, 1998

It is also true that the focus of mining is the extraction of non renewable resources. Ore, once removed from the ground cannot be regenerated. When mines are closed, their direct economic value comes to an end often with traumatic economic and social implications for those affected, as the "ghost town phenomenon" so aptly illustrates. Finally, mining disturbs the natural environment. Where open pit mining is involved, the disturbance can be massive. The Syncrude operations in the Athabasca Tar Sands move 300,000 tons of sand a day. The Bingham Canyon copper mine in Utah (USA) is the largest human excavation in the world involving the removal of more than seven times the amount of material excavated to create the Panama Canal (Young, 1992). Underground mining leaves a much smaller environmental "footprint" it is true. Yet underground operations too can have significant impacts, a fact well illustrated in mining centres like Sudbury (Canada). Hence mining is a relatively easy target for environmental critics and for communities looking for stability and wellbeing into a future they want to continue beyond the life time of an ore body whatever the short term prosperity it might bring.

Is mining, then, incompatible with sustainable development? There is much evidence to suggest that it is not, as the presentations of other members of this panel will no doubt show. Is the concept of sustainable development anything more than a Trojan horse to be regarded justifiably with the suspicion that it frequently encounters in mining circles? I think not. Rather, it is a concept that should be carefully explored and applied. The benefits of so doing, my research suggests, are substantial.

THE ANATOMY OF A RESEARCH PROJECT

It is widely assumed that environmental conflict rests on deeply divergent environmental, social and economic values. Conflicts generated by mining are no exception. Further, whether mining or other forms of resource extraction are involved, this assumption can have very discouraging implications. It seems to suggest to many people, including often the participants themselves, that the conflicts in question cannot be mediated or resolved through dialogue or debate grounded on reliable research and scientific analysis. This in turn implies a need on the part of combatants to mobilize forces, exert political pressure using whatever methods it takes and, if all else fails, resort to various forms of civil disobedience on the one hand or dubious political interventions (bribery for example) on the other. Yet at the same time, those involved invariably think their own positions to be rationally defensible. That is why they are willing often to go to extreme lengths to defend them. But why, if that is the case, is conflict and political manoeuvring so often seen as necessary and justifiable?

At least two elements are basic to these assumptions about the nature of environmental conflict. First is the assumption that the values being put in play by the participants are radically different. Second is the assumption that those values rest on incompatible environmental perspectives or environmental philosophies. These assumptions are fundamental to the interpretation of environmental conflict. A York University based research project set out to test them. The result is a detailed and systematic study of four cases. And though only one of them takes mining as its focus, none the less, the findings emerging from this study have direct application to the issues with which we are here concerned.

Four Case Studies

The first of our cases is a timber management plan that governs the use of the timber resources in a broad area south and east of Sudbury in mid-northern Ontario. The plan was successfully negotiated with wide stakeholder involvement earlier this decade. Access to the timber is of importance to several forest products companies operating in that area of the province. However, the land in question is also used extensively for recreation purposes: camping, canoeing, hiking, cottaging, fishing and hunting. All of these uses involve commercial, private, and public interests which in this case were involved in the negotiation process which led earlier this decade to a new five year plan..

The second case takes as its focus the use of the timber resources in Algonquin Park, Canada's oldest provincial park. The area has a rich natural, recreational and commercial history. It is a wildlife preserve, has an interesting and varied geology and geography and is a popular destination for campers, cottagers, canoeists, hikers, bird watchers, wild life enthusiasts and scientists. It is also an important source of soft and hardwood lumber and has been the scene of intensive logging for more than one hundred years. Finally, the area is the subject of a land claim by a First Nation band whose members have only just recently been allowed by the provincial government to resume limited hunting within the park boundaries. Hence, as with the Trout Lake area, private, commercial and public interests are involved. The timber management plan which governs logging in the part has until recently attracted wide agreement. However, it is now being challenged by a group that wishes to see logging prohibited. Furthermore, the First Nation band resident on the eastern boundaries of the park are claiming traditional use rights including hunting and fishing and presumably the exploitation of the park's other natural resources. Hence it introduces issues associated with aboriginal rights that are not infrequently encountered by mining companies wishing to exploit ore bodies discovered in areas in which there are aboriginal residents.

Our third case is a proposed open pit copper mine development planned for northern British Columbia by Placer Dome, a multi national Canadian mining company. The issues this case raises are characteristic of mining developments: the potential for ground water pollution from tailings sites, the opening of wilderness areas through road construction, wildlife habitat destruction and so on. Prior to developing its proposal, Placer Dome identified possible stakeholders including two First Nation Bands and invited their participation in the planning process. The resulting plan has subsequently won government approval following extensive environmental review by the stakeholders involved.

Our last case involves the harnessing of water draining from Moose River Basin into James Bay. This is the most complex of our studies because of the wide area affected, the number of stakeholders and the importance of the interests at stake. Until recently the proposed hydro electric developments were a part of a Twenty-five Year Supply Demand Plan developed by Ontario Hydro and subsequently submitted for review under Ontario's Environmental Assessment Act. Over $50 million were disbursed to interested parties to finance their evaluation of the plan in preparation for environmental assessment hearings. The twenty five year plan was withdrawn by Ontario Hydro in light of dramatic reductions in the demand for power earlier this decade. However, the documents generated by stakeholder coalitions participating in the environmental assessment process were largely completed before the plan was withdrawn. Although the larger plan was abandoned, Ontario Hydro decided to continue a relatively modest redevelopment of four dams on the Mattagami River, a major river draining into James Bay. Stakeholders in the project include a First Nations coalition, the Ontario public, several government ministries, forest companies, several municipalities, hunters, trappers, and several environmental groups to name just a few.

The four cases were chosen for two reasons. First, they involved conflict of a non-politicized nature. That is to say, cases (like for example Clayoquot Sound or Temagami) were avoided where positions were likely to be so deeply entrenched that responses to the research agenda were likely to be formulated largely in terms of their anticipated rhetorical impact on a public audience. Equally important was to avoid situations where it would be difficult not to become a party to the conflict. Finally, it was important to study cases that had been through some kind of public environmental assessment that had left a substantial paper trail. The four cases fitted these requirements.

Research has been conducted in two phases. The first phase consisted of site visits and document analysis. The purpose of the site visits was to meet stakeholders and familiarize ourselves with the cases in a concrete way. Document analysis was undertaken by graduate research assistants at York University drawn from business, environmental studies and philosophy. A complex coding scheme was constructed whose purpose was to allow us to build a values inventory and explore the logic of the values in that inventory.

The second phase involved interviews with case stakeholders. Over eighty interviews were completed. Our goal was to interview five people from each of the groups having a direct stake in a case under study. People interviewed were asked first to respond to questions about the case in which they had a direct stake. They were then asked to address questions about the other three cases under study. Interviews were completed by the spring of 1996.

Relevant Findings

What the case studies revealed was four different levels of conflict and cooperation, trust and mistrust and respect and disrespect across a continuum. Analysis of the stakeholder documents and interviews suggests that these varying levels of conflict, trust and respect were correlated with two factors of particular interest for this discussion. The first was the clarity, honesty, transparency and defensibility with which resource use project proponents succeeded in setting out their core values. The second was the inclusion of sustainable development as one of those core values.

Although many resource extraction companies, guided as they typically are by the vocabulary of engineering, are not intuitively comfortable with values-based terminology, none the less, values-based terminology has come to play a central role in environmental planning and in environmental debates and conflicts. Thus it is not at all uncommon to encounter discussions of "forest values", "park values", and "recreational values", to provide just three common examples. What is more, the language of values is helpful in identifying key factors in environmental planning and resources use evaluation and criticism. For values play a central role in the planning and the managing of corporations just as they do in the decisions and choices of individuals. They guide the selection of priorities. They are revealed in preferences, words and actions. They help to determine goals and objectives. To return to language more familiar to mining, they also guide the identification and calculation of costs and benefits.[1]

Core values have to do with fundamental goals and objectives. They constitute people's reasons for doing whatever they are doing. Getting clear on core values is a basic function and a key responsibility of management.

Why is clarity, honesty and transparency in setting our core values crucial to facilitating communication and building trust and respect? The answer is relatively easy to discern, at least in its broad complexion. The usual view of resource use companies, including of course mining companies, is that their development proposals and their approach to management is fundamentally self-serving and oriented to profit maximization.[2] What follows from this view, in the minds of many observers, is that the concern of resource extraction companies for other stakeholders and for the environment is inevitably minimal. This in turn has significant spin-off implications. It means, for example, that evidence produced by resource extraction companies to demonstrate the benign nature of projects being proposed or underway, or the capacity to reduce environmental risks is likely to be greeted with considerable scepticism. This distrust is frequently carried over to include scientific research supporting the company position which is therefore greeted as untrustworthy particularly when the research is financed by the company under scrutiny. This kind of perspective is reinforced in the public mind by pictures of abandoned mine sites, unsightly tailings ponds and slag heaps. Crucial to countering these optics is an clear and defensible account of the core values designed to guide mine development and on-going mining operations. Of course, what constitutes a "defensible" set of core values is certain to be contested. Nevertheless, it is an important first step.

In two of the four cases we studied, the Trout Lake Timber Management Plan and the Ontario Hydro Mattagami Extension project, failure to build project proposals around clear and defensible core values played a key role in the hearing granted both projects. This is not to say that in these cases, project proponents failed to identify the core values around which their goals and objectives were organized. What was lacking was the element of defensibility. In each of these two cases, a key stakeholder group failed to be convinced that the values articulated were the values that would in fact guide resource extraction or use.

In the case of Ontario Hydro, there were two problems. First, the crown corporation identified their goals and objectives not as there own but rather as the goals and objectives of the "people of Ontario". That is to say, the way in which Ontario Hydro articulated its core values in its own environmental assessment of the Mattagami Extension project, implied that its most fundamental organizational commitment and guiding value was to respect the values that in the companies view the people of Ontario wanted respected in the expansion of hydro electric generating facilities. This might seem an obvious and commendable objective for a crown corporation, particularly in light of Ontario Hydro's understanding of the values that the people of Ontario wanted the crown corporation to respect.[3] But was this stance defensible in light of what was being proposed? Arguably it was not in as much as it communicated the message to those likely to be impacted by the development that if the wishes of the people of Ontario -- as interpreted by Ontario Hydro! -- were to shift, so would the corporation's understanding of its goals and objectives and therefore its obligations to project stakeholders. To concerned external stakeholders, this did not look like a long term commitment regardless of the attractiveness of the goals and objectives in question. Political winds shift, sometimes quite quickly. And what people want from their political masters and servants is subject to varying political interpretations.[4]

The uncertainty generated by this way of articulating the corporation core values was exacerbated by a second important factor, namely the history of hydro electric power development in northern Ontario. Ontario Hydro's legislated mandate throughout most of the development period was: "To provide a reliable supply of electrical power and energy to the people of Ontario at the lowest feasible long-term costs" (Environmental Assessment Summary, p. 3). The province and the people of Ontario have been the direct and significant beneficiaries of that mandate. The people of northern Ontario, particularly the north's aboriginal inhabitants have not. The reasons are not hard to discern. For most of this century, Ontario Hydro interpreted costs to mean financial costs. Their strategy was "least cost planning", an approach common to utilities throughout North America throughout this century (Litchfield, 1994). The problem is that this strategy both required and justified the externalization of costs where the law allowed. The externalized costs were carried by the local communities, which were, as indicated above, largely aboriginal.

It is important to note here that this planning strategy was not confined just to public utilities like Ontario Hydro. It was in fact the general rule that guided resource projects for the most part throughout the north, including, of course, mineral extraction. The result has been a legacy of grievances and mistrust. It was this legacy that dominated the response to Ontario Hydro's plans for redeveloping the hydro electric potential of the Mattagami River. Ontario Hydro did not deny it had an obligation to respond to these past grievances. But it did refuse to respond to the demand that it address the issue of past grievances within the parameters of the environmental assessment process. What this implied is that for Ontario Hydro, assessment of the environmental acceptability of a project was to be determined independently of past conduct. What this stance failed to recognize was that track record is one of the filters that project stakeholders use to interpret things like sincerity, trustworthiness, commitment and so on. The consequence was that the high sounding values Ontario Hydro claimed would guide the redevelopment project were regarded with deep scepticism particularly by the aboriginal communities directly involved.

The second case exhibiting serious conflict was the Trout Lake Timber Management Plan. Here too the core goals and objectives of timber management were mandated by provincial legislation. The purpose of the plan was to put those goals and objectives into operation. As with Ontario Hydro, the core values around which the plan was developed were clear and on the surface at least defensible[5]. However, also as with Ontario Hydro, the plan articulated by the key Ministry of Natural Resources officials was greeted with deep scepticism. What lay behind the conflict was the belief that the economic and the environmental objectives being advocated were incompatible. Hence, the local campers association was of the view that when it came time to implement the plan, environmental protection would be sacrificed to ensuring a continuing adequate supply of logs for the local forest products industry. Our interviews with ministry officials suggested, furthermore, that these fears were not entirely imaginary.

These findings suggest a first conclusion. A clear, honest, transparent statement of core values is important because it is a vital first step toward establishing accountability, credibility and therefore defensibility. To put it another way, resource extraction is not going to be seen as defensible unless the guiding values can be identified and tested. What is more, if they are not articulated by project proponents, a substitute set will be projected onto project proposals by sceptical external stakeholders. However, there is a second important lesson to be learned from the discussion thus far as well. It is unlikely that project proposals are going to be seen as defensible unless the development they call for is sustainable. This was a key factor in the conflict generated by the Trout Lake Timber Management Plan and the Mattagami redevelopment proposal advocated by Ontario Hydro. Although the reasons for worrying about the sustainability of the resource use envisioned in these two cases were different, the underlying worry was similar. In both cases, a group that saw itself as having a great deal at stake had serious doubts about negative long term implications of the proposal under evaluation.

Sustainable Development as a Core Value for Resource Use Industries

Seen from a very narrow environmental perspective, the concept of sustainable development applied to mining is an oxymoron. [6] Without either denying or disguising the fact that mining has to do with the extraction and use of non renewable resources, it is none the less an industrial activity for which principles of sustainable development have direct and meaningful application. What is essential, however, is that those who appeal to it or accept it as a core guiding value recognize that the concept is a complex one. When its components are set out, its relevance for the future of mining become clear. Those components, I suggest, are four in number.

i) Most obviously, sustainable development must include environmental conservation and protection. A first key question for mining is the extent to which it can be carried out without creating "down stream" pollution. The effect of pollution is to reduce or cancel the value of the natural or human environment affected seen from the perspective of other users including non humans. A second key question is the extent to which an area which is mined can be returned to a natural condition. What counts as a natural conditions will be a subject for dialogue with affected stakeholders. Clearly, it will not include returning the land to its pre-mining condition. What it will include, however, is land rehabilitation with a view to reintegrating the land in question with its surrounding natural environment. This in turn will require scientific study and dialogue sometimes of a protracted nature with other stakeholders. Finally, a commitment to environmental conservation and protection will require that the industry as a whole and individual mining companies in particular recognize that some deposits, perhaps even very rich ones, will have to be left undisturbed because mining them is incompatible with that perspective. This might be the case where the land in question had overriding cultural significance as a heritage sight, for example a national park like Banff or Yellow stone. Or developing the land might pose serious pollution risks because, for example, the deposit was located in an area prone to earthquakes.

ii) A less frequently recognized component of sustainable development is its ethical dimensions[7]. Sustainable development requires that the costs and benefits be accounted for and shared fairly. This in turn requires two things. Sustainable development requires first that mining companies identify honesty and fairly all those who are likely to be directly affected by the development of a mine. The more obvious of these are employees, investors, suppliers and buyers. The less obvious include local communities, families of employees and others who might be put a risk in some way by the development as a result of its social, economic or environmental impacts. Second, costs and benefits must be determined from the point of view of all a project's stakeholders and not just its shareholders or those with political influence, for example. This poses a real challenge for mining. The problem lies not in the idea of cost/benefit analysis which is no stranger to mining company operations. Rather, the problem lies in the fact that traditional approaches to determining costs and benefits bring with them certain hazards. One of these is the assumption that a monetary value can be put on all costs; or alternatively those costs that cannot be monetized can be safely or legitimately ignored. Both of these assumptions are faulty. One of the most serious mistakes that a development proponent can make is to offer "market value substitutes or financial (e.g. market value) compensation where a substitute is not available to an individual or group for something whose value from their perspective cannot be measured in financial terms. Examples are activities or things that symbolize or are an integral part of a way of life.[8] A second hazard is an assumption closely related to the first two and frequently attached to cost/benefit analysis. This is the assumption that cost/benefit analysis is culturally neutral and a matter for experts.[9] However, if not all costs are quantifiable via market value comparisons, then, identifying costs and assessing the value of benefits will require careful dialogue with project stakeholders. In the absence of that dialogue, projects may well be incompatible with sustainable development objectives.

Finally, sustainable development requires that costs not be imposed which those on whom they are imposed are unlikely to recover from or alternatively from which they can recover only with great difficulty. Mining operations that pose serious risks to the health of employees or neighbouring communities would count in this respect as would threats to natural environments that are national heritages or irreplaceable for social or economic reasons or a habitat for endangered species.

All these considerations are ethical in nature because they involve the fair distribution of costs and benefits flowing from resource extraction or use.

iii) A third component of sustainable development is efficient and effective resource extraction. This is not always acknowledged in discussions of sustainable development. This is because many people view efficiency as a purely economic value linked directly to profit maximization. In the minds or many, this puts efficiency in direct conflict with environmental protection and the pursuit of social justice. This view is grounded on a misunderstanding. Efficiency has to do with accomplishing goals and objectives without causing the avoidable loss or waste of things of value. The efficient use of resources will maximizing environmental and social as well as economic benefits as described in (ii) above while minimizing costs also defined broadly to include negative environmental and social impacts. Here the operative maxim is "waste not want not". It connects with environmental conservation and protection. Inefficient use of a resource is one way of ignoring principles of conservation. It involves failing to acknowledge the total value of a resource seen from the perspective of all those with an interest in it. Inefficiency therefore undermines the fair distribution of costs and benefits both for those who would otherwise derive benefit in the present from a more efficient use of the resource in question and intergenerationally. Inefficient use of present resources leaves less by way of natural resources to be handed to the next generation while storing up potential costs associated with the possible future need for rehabilitation of land for example or waste disposal.

An alternative way of putting this is to point out that externalizing costs associated with a mine is a form of inefficiency. It implies that something which is of value has no value and can therefore be treated as such. One of the key challenges of sustainable development is therefore to identify the value of natural resources in accurate and insightful ways and then to extract those resources in ways that does not needlessly destroy the value of what is "down stream" or what is left behind both for present and future generations.

iv) A fourth component of sustainable development is accountability. Accountability enhances trust. It allows stakeholders to evaluate performance against public commitment to the principles of sustainable development put in place by resource extraction industries or resource extraction companies. Commitment to sustainable development provides opportunities for modelling best practice and invites stakeholder cooperation and participation in its achievement. It helps to create a social environment in which sound research is supported, shared and respected thus enhancing dialogue rather than denunciation and consensus building rather than conflict and confrontation.

The significance of these principles was evident in the cases we studied. Two of our cases were at the low end of the conflict continuum. The Mount Milligan project generated less conflict for several reasons. The most significant trust builder was the company's very strong stance on environmental protection, a stance which an extended process of consultation allowed the company to communicate to local community stakeholders. In this case, the company committed itself to best practice technology and research with a view to ensuring that mining and milling processes would not result in the pollution of a creek running through the property thereby protecting a river, the Nation River, for which the creek was a tributary.

An important component in the trust building process was a strong framework of environmental laws enacted by the provincial legislature. Our interviews indicated that the lack of deeply entrenched opposition to the project was grounded on a belief not that the company could be trusted in the absence of any other restraints to protect the environment. Rather, it was based on the belief, which was apparently quite widely shared, in the efficacy and effectiveness of the prevailing environmental protection laws combined with a conviction that the company would respect those laws. Openness to involving the local community in environmental monitoring was a factor as well.

Equally interesting in this regard was the fact that at least some senior managers viewed the law in a similar light. That is to say, they saw the environmental protection laws as welcome "obstacles" that the company had to clear if the mine was to become a reality. Not all the management personnel we interviewed were convinced that the traditional profit maximization orientation of the business world would allow environmental protection as a high priority independently of legislated standards. Since they as individuals believed environmental protection to be important, they welcomed the law as a kind of insurance policy requiring that the company live up to the high environmental standards to which it was publicly committed.

Equally revealing was the fact that the level of confidence in the company's commitment to environmental protection among First Nation leaders was less pronounced. Here too the stance that emerged was connected to the law. The First Nation leaders we interviewed seemed confident that the company would respect both the letter and the spirit of the law. But they had less confidence that the law offered a long term guarantee. That is to say, they were in effect more sensitive to the possibility that legal standards might be watered down by subsequent governments. Hence legal guarantees were less reassuring to people whose sense of obligation to future generations was rooted in moral and spiritual values connoting respect for "mother earth".

Also a factor in building trust was the fact that principles of economic and social sustainability were also built into project planning and explored at length in dialogue with the projects external stakeholders. The company acknowledged publicly an obligation to support the development of skills and business opportunities whose value would not be lost when the mine closed. (The projected mine life was about ten years.) Social sustainability beyond the mine's projected life time was also addressed. Interviews indicated that the community opinion leaders were persuaded that they could expect cooperation from the company in avoiding "boom and bust" economic impacts. There was less confidence about social impacts. Here the thinking the company brought to the planning process was more conventional and less convincing. None the less, community leaders were reasonably optimistic that problems would be manageable with the cooperation of mine management.

The same was not true of the aboriginal community. Here the conventional character of company thinking about social sustainability was least convincing. The two adjacent aboriginal bands were concerned about the impact of high wages on their young people and on community values. They anticipated socially disintegrative impacts on their young people. Access to wage labour was also perceived as a threat to traditional subsistence values. Would respect for the land withstand the pressures of a wage economy? How could those traditional values be supported and encouraged? The remedies proposed by the company were those that had evolved in response to the demands of conventional urban patterns of life, substance abuse counselling, for example. The perceived need on the part of the aboriginal community was for more community based, communal support systems. It was clear from the documents available to the research team and the subsequent interviews, that the company was uncomfortable about the precedents that might result from departure from traditional understandings of social obligations. The result was a pervasive ambivalence about the project on the part of aboriginal stakeholders.

Analysis of the debate and dialogue generated by the Algonquin Park Forest Management Plan indicated the lowest level of conflict of the four cases under study. There were several lessons to be learned from this case. The core mandate was again a legislated one. However, sustainable development of the Park's forest resources was at the heart of that mandate. More important, it was clear that the Algonquin Forest Management Authority was committed as an organization to sustainable forestry. Most important of all, analysis revealed that the Authority had succeeded in communicating that commitment to its various stakeholders. It is interesting to note here the core values around which the plan was developed. The highest priority was the long term economic health of the region. This was understood to require both logging and tourism both of which were seen as building on effective forest and land management. A second core value was protecting opportunities for wilderness experience on the part of park visitors. Both these values were understood by the Algonquin Forest Authority to rest on environmental protection and conservation.

One of the most interesting features of the case lay in the fact that a key environmental group active in the park was deeply opposed to continued logging within the park boundaries. They argued that protecting wilderness values should take the highest priority and that wilderness values were incompatible with continued logging. Their views were well known as the group was a persistent public advocate of this position. What is instructive about the case, however, is that this significant and deep disagreement about the environmental values that should govern use of the Park and its forest resources did not generate serious conflict. To the contrary, although the environmental group in question opposed logging and continues to do so, they respect the way in which the Algonquin Park Authority is implementing its mandated responsibility for managing the Park's forest resources.

The basis for this respect is the obvious commitment on the part of the Authority and its staff to sustainable development values. The Authority's environmental critics do not share the Authority's concept of sustainable forestry. In their view, the most important Park values can only be sustained through a termination of logging. On the other hand, they accept that the Authority's commitment to sustainable logging is sincere and that their approach to forest management, given their concept of sustainability, is exemplary and a model of how logging should be carried on where it is permitted. In a word, exchanges between the "antagonists" been marked by respect on the part of the environmentalists. Equally, the Authority recognizes the importance of ensuring the efficient use of the resources under their oversight understood in as efficiency is described above. They have welcomed the processes in place to encourage both communication and accountability. They regard the environmentalists as partners and welcome their input even though it is clear that the Authority itself disagrees with the view that logging should be eliminated within the Park.

What our research also showed was that the local logging industry people interviewed supported the logging values espoused by the Authority and believed that the logging firms working in the Park had been won over to the need to incorporate principles of sustainable development into their logging practices both inside and outside the Park by the Algonquin Park Forest Authority.

In conclusion, it would be incorrect to suggest that the management of timber resources of Algonquin Park was or is without conflict. It is fair to conclude, however, that the disagreements involved have not undermined respect for the position and work of the Park's Forest Authority.

A final note should be added with respect to the First Nation community living outside the Park boundary. This community had a long standing land claim that encompassed the Park as well as much of the land surrounding it. Because of the nature of its land claim, the First Nation Band involved refused to participate in a formal way in the Forest Management Plan assessment process. They were also the only group that refused to be interviewed as a part of the research project. Nevertheless, informal dialogue with the Band's chief together with a study of extensive documentation that the Band had developed to justify and explain their claim to the public generally indicated again underlying respect for the management practices of the Algonquin Park Forest Authority. The fundamental disagreement here had to do with who should manage the Park and not how it should be managed.

CONCLUDING OBSERVATIONS

I began this paper with two observations: first articulating a clear and defensible vision of the values that lie at the heart of resource extraction projects is crucial to building public confidence in mining and the mining industry ; second, commitment to principles of sustainable development has an important role to play in building the basis for communication, reducing conflict and building public confidence and trust in companies who espouse them who are engaged in mining. I have not demonstrated in a detailed way that mining is compatible with sustainable development. That is a task for others on the panel and for detailed analysis at another time. What I have defended and illustrated is the importance of these two observations for the future of the mining industry. If mining stakeholders are not presented with clear and defensible statements of core values, they will substitute their own views of what mining companies are really after. And those visions will build around traditional and largely discredited factors such as profit maximization understood as requiring the externalization of environmental, economic and social costs on a naive and exploitable public. In the absence of sustainable development principles at the core of mine planning, design, operation and decommissioning, the values guiding mining as an industrial activity will not be seen as defensible or credible over the long term by that same public. These two requirements therefore constitute the way ahead and the most significant challenge for the mining industry as it moves into the next century.

---

[1] For an interesting though complex exploration of the role of values in cost benefit analysis, see Brunk (1991).

[2] That such a view should be projected on management is hardly surprising since this perspective dominates management education in North America as is suggested by Neil Shankman in paper in to be published in The Journal of Business Ethics (Shankman, forthcoming)

[3] Those values were in order of priority: *social acceptability* interpreted to include heritage protection, low cost development, and efficient use; and *environmental acceptability* interpreted to include environmental protection and conservation;

[4] A fact which is born out by political shifts that have occurred in this decade in countries like Canada as reflected in the changing priorities of provincial and federal governments on environment issues and the priority to be given to environmental protection.

[5] The core values of the Timber Management Plan were: *the long term economic welfare of society* which was interpreted to mandate the use of the forests covered by the planning process to support the local forest products industry and the jobs tied to it; *outdoor recreation* which required a high value being placed on fish, water quality, wildlife; and *environmental protection.*

[6] It is true, of course, that there are environmentalists who would claim that this is true for virtually any form of development as that term is understood in advanced industrial countries. In this sense, mining is no different from any other kind of modern industrial activity.

[7] The dimensions of this component are explored in considerably more detail in two articles: "Ethics Surface Mining and the Environment" (Cragg, 1995) and "Sustainable Development and Historical Injustice: Lessons from the Moose River Basin (Cragg, 1996).

[8] There are many examples in the history of resource extraction projects that illustrate this mistake and the costs attached to it by way of hostility and mistrust. At a recent seminar involving partners in the "four case study research project" on which this paper is based, an aboriginal participant described the feeling of anger and deceit that was generated by resource project that destroyed the fish in a previously salmon rich river on which the local aboriginals had traditionally relied. The company offered to truck in fish which they proposed to distribute at no cost to those affected. While this did not constitute financial compensation it was a "market value" substitute. The effect of the gesture was to generate real bitterness and an entrenched sense of grievance. For what the company has done was to alter fundamentally a way of life built around subsistence fishing. The fish had a market value; the role of fish in the culture did not. To suggest it did was simply to add insult to injury.

[9] See note 1. above.

****

REFERENCES

Brunk, C., Haworth, L. and Lee, B. 1991.
Value Assumptions in Risk Assessment: A Case Study of the Alachlor Controversy, Wilfred Laurier University Press, Waterloo, Canada.

Cragg, W., Pearson, D. and Cooney, J., 1996.
"Ethics, Surface Mining and the Environment, Resource Policy, Vol. 17, No. 10, December.

Cragg, W., and Schwartz, M., 1996.
Historical Injustice and Sustainable Development: Lessons from the Moose River Basin, Journal of Canadian Studies, Vol. 31, No. 1, (Spring).

Dungan, P., 1997
Rock Solid: The Impact of the Mining and Primary Metals Industries on the Canadian Economy, Institute for Policy Analysis, University of Toronto, Toronto, Canada.

Young, J.E. 1992.
Mining the Earth in L.R. Brown (ed.) The State of the World 1992, W.W. Norton, New York.

The future role of research in the mining and metal processing industry

I.J. Itzkovitch
Noranda Technology Centre, 240 Hymus Blvd.
Pointe-Claire, Québec, Canada H9R 1G5

ABSTRACT

The role of research in the mining and metal processing industry is to provide new and better ways of producing, using and recycling metals. Environmental research is an integral part of this process as are the investigation of related technical, economic and social parameters. This presentation looks at research in the broader context of contributing to the improved performance of the entire enterprise with benefits to society at large. This is the most cost effective way of achieving environmental improvements. This presentation identifies some of the needs and opportunities for intensified research in the Mining and Metal Processing Industry and discusses some of the ways in which this need may be met in an expeditious and cost effective way.

Proceedings of
Metals and the Environment: An International Symposium
Edited by Jim Skeaff
The Metallurgical Society of CIM
100th CIM Annual General Meeting, 1998

INTRODUCTION

The role of research in the mining and metal processing industry is to provide new and better ways of producing, using and recycling metals. Environmental research is an integral part of this process as are the investigation of related technical, economic and social parameters. This presentation looks at research in the broader context of contributing to the improved performance of the entire enterprise with benefits to society at large. This is the most cost effective way of achieving environmental improvements. This presentation identifies some of the needs and opportunities for intensified research in the Mining and Metal Processing Industry and discusses some of the ways in which this need may be met in an expeditious and cost effective way.

DEFINITION

Webster defines research in noun form as a careful search or investigation; systematic investigation towards increasing the sum of human knowledge. Research requires a thorough study of what has been done previously to avoid 'reinventing the wheel'. Thus a better definition is perhaps the verb *to re-search*' or 'to look again or reexamine'.

RESEARCH TARGETS

Any systematic search of facts for enlightenment can be considered research. Industrial research is constrained in that the objects of study need to offer some reasonable prospect of ultimate economic benefit to the organization and is often driven by immediate economic need. This limitation still leaves a very broad range of targets for systematic and economically productive investigation including those listed in Table 1.

Table 1: Research Targets

Production Related	Community Related
exploration	**energy efficiency**
mining	**ecological interactions**
milling	**industrial hygiene**
concentration	**health effects**
transportation	**risk management**
smelting	**forensic audits**
refining	**impact assessments**
marketing	**human relations**
by-product use	**community perspectives**
product design	**public relations**
recycling	**employee attitudes**

There is clearly a multitude of opportunities for productive research. The list of possibilities is endless. Careful selection of research targets is vital if best use is to be made of people, facilities and funds.

RESEARCH MANAGEMENT

Industrial research is directed to invention. Invention has been defined as 10% inspiration and 90% perspiration. It is hard work and it is expensive. The risks are great as also are the rewards for success. Good management can minimize the risks without losing the rewards. Table 2 lists some actions often identified with good research management.

Table 2. Some Guidelines to Sound Research Management

Guideline	Requirement
target selection	criteria development
literature study	past work
patents review	related concepts
existing systems analysis	key element identification
operations visits	field assessments
team selection	competence & motivation
research facilities	suitability for work
piloting options	speed versus risk evaluation
third party assessments	search for fatal flaws
engineering linkages	technology transfer
financial risk analysis	economic sensitivity
environmental assessment	ecological interactions
discharge characteristics	regulatory internal objectives
permitting needs	process auditing
social assessment	community consultation
start up crew training	adequate preparation

Good research management requires attention to all the factors listed in Table 2 and many more.

TEMPORAL CONSIDERATIONS

Existing operations frequently have a life of 25-50 years. Continually incremental improvement effort over this life time results in major aggregate improvement in performance. New facilities, which are constructed relatively infrequently to meet new needs or opportunities, build on this accumulated knowledge and experience. Elements of technology are combined from many existing plants in new and novel ways with a leavening of new technology developed through laboratory investigations and pilot operations. The new facility also provides an opportunity to benefit from other technologies such as design and construction techniques, new materials, new equipment and instrumentation automation. These parallel developments can permit operation in ways that were previously impractical or uneconomic.

RESEARCH COOPERATION

There are many examples of cooperative industrial research in the recent past. Some of the programs involving the mining industry either as a group or as individual companies include those shown in Table 3.

Table 3:
Examples of Cooperative Research Between Industry, Government, Academia & Others

Name	Function	Acronym
Mine Environment Neutral Discharge	**AMD research**	**MEND**
Whitehorse Mining Initiative	**Societal links**	**WMI**
Queen's Centre for Resource Studies (past)	**Economics**	**CRS**
Aquatic Effects Toxicity Evaluation	**Protocols**	**AETE**
Thiosalt, Jarosite & Slag Consortia	**Chemistry**	**None**
Centre Res'rch Exc'll'nce in Terr'str'l/Space Tech.	**University**	**Crestech**
Canadian Institute for Advanced Research	**Geological**	**CIAR**
Environ'l Science & Technology Alliance of Canada	**Linking**	**Estac**

The International Council for Metals and the Environmental has sponsored special technology review studies on environmental and health topics including recycling and toxic effects of metals. Other associations have also sponsored special studies either individually or as a group.

FOCUS OF CURRENT RESEARCH EFFORTS

The focus of current research efforts are summarized in Table 4.

Each of these fields give rise to a multitude of questions and concerns, ideas and opportunities which could easily consume the entire revenue of any organization short of the US Federal government and it is critical to be selective and focused in using limited resources. The challenges related to each of these subjects are discussed in more detail below.

Table 4: Mine Environment Research Fields

Metal Mining & Processing Research Scope
1. Exploration and Mine Technique Improvement
2. Clean Technology Development
3. Air Emission & Water Effluent Control
4. Solid By-Products Utilization
5. Metal Use, Recycling, Reuse and Dispersion Management
6. Energy Efficiency & CO_2 Reduction Projects
7. Biotechnology for Metal Processing & Remediation

DISCUSSION OF RESEARCH CHALLENGES

1. Exploration and Mine Development

The search for new potential orebodies is a research activity applied to the top kilometre of the earth's crust. There is a need to improve exploration techniques both to reduce the cost of locating new ore and to find potential ore at greater depths. Mine development is at present limited to about the top two kilometres of the earth's crust although there is little doubt that the mineralized zones under many mines goes much deeper and ore reserves would be greatly expanded if it were possible to economically extract such metal bearing material mineral. There is therefore great scope for research directed to the finding and extracting ore at depth.

2. Clean Technology and Best Available Technology (BAT)

Mining and processing intrude on the natural world as do all human activities. It is necessary to be vigilant in understanding and assessing the risk and benefits associated with all activities and where necessary mitigate the risk. Existing plants and processes, even with the benefit of continual improvement programs, have scope for further improvement. There is also a need to develop new processes to meet the anticipated economic, social and environmental requirements of 10, 20 or 30 years hence. The new generation of processing facilities are required. These must be less costly to build and operate, provide exemplary working conditions and incorporate prevention as opposed to control of environmental impact. This requires attention to design of both the process and plant facilities. Since each ore is different, each mining and processing operation has to be different in detail. The concept of Best Available Technology for each metal is a myth in that it fails to recognize and accommodate the need for diversity and flexibility in design. Legislated BAT fossilizes inferior technology and mandates use of inappropriate technology. It is counter-productive. A more productive approach is to capture detail of, and critique current technology for, reference purposes in a guidance document, manual or handbook (hard copy or electronic form). BAT is and needs to be a constantly moving target as new information becomes available and is incorporated into new or existing plants. Higher metal demand and prices provide the incentive to develop new operations incorporating the best technology available. New operations are usually the most efficient and profitable. Adequate and persistent research effort provides the new technology.

3. Air & Water Use

Mines and processing plants also use large volumes of water and consequently treat large volumes of effluent to remove undesirable components, producing a clean product suitable for discharge or, increasingly, recycle. Smelters and to a lesser extent mines, mills and refineries use large volumes of air for processing and discharge the resulting gases to atmosphere either via stacks or roof fans and monitors. Major progress has been made in recent years in reducing emissions from all processing facilities. Gases prior to stack discharge are usually treated to remove the majority of the sulphur dioxide for conversion to sulphuric acid and the majority of the dust for recycle to the process. There is continued emphasis on incremental process improvement in existing plants as some sulphur and some dust still escape and a continued focus is required to find ways of economically removing more. There is a need in the longer term to find a radical way of decoupling the processing of sulphides from the production of sulphuric acid.

4. Solid By-Products

The main product of mining and base metal processing in tonnage terms is solid waste. Much research effort has been directed to isolating solid wastes effectively, particularly wastes containing potentially reactive sulphides, with considerable success. Some of this waste in the form of crushed rock or deslimed tailings is placed underground to provide ground support, the rest accumulates as piles on surface which are usually covered with water and/or inert solids before or after closure of the mining operation. While it may be legitimately claimed that these are stores of values for future generations, the cost of preserving this material is substantial and it would be far preferable to find some profitable and ecologically acceptable use for the material as it is produced. Since we are dealing with large tonnages of low value commodities this is a tough challenge. The short term prospect for economically marketing by-products is primarily in the niche market. Research to tailor by-product materials to meet the need of niche markets can be beneficial.

5. Metal Use, Recycling, Reuse and Dispersion

Metals are vital elements in any industrial economy and with expansion of the global population and rise in living standards the demand for metals is increasing. Table 5 shows some major metal uses.

Table 5: Major Metal Uses

Metal	Property	Illustrative Use
Copper	High conductivity	Power transmission
Aluminum	Low weight	Truck bodies
Nickel	Ease of fabrication	Stainless steel products
Lead	Stability	Rechargeable batteries
Zinc	Electrochemical potential	Steel rust protection

The expanding needs of the expanding world population for metals has to be met if quality of life is to be improved. The intensity of metal use is progressively increasing. The total demand for metal still exceeds the availability of recycle metal. It is therefore unrealistic as some suggest to eliminate primary production and survive on recycled metals Meanwhile work continues to find ways to make more efficient use of metals and improve recycling technology.

Finding acceptable and beneficial ways to reuse materials with low concentrations of metals is another industry challenge. Examples are metal species that have become diluted as in dusts, sewage, grindings, ash, slag and similar fine materials. The cost and energy for transportation and processing may be too high to warrant extraction of pure metal and in this case alternative uses for the bulk commodity are sought. Some of these products are beneficially added to cement kilns to improve cement properties, to fertilizer to provide trace nutrients required by plants or animals for health, used as bulk fill or incorporated in aggregate. More research effort to find ways to reduce dispersion and to find such beneficial uses is required. Life cycle analysis can help to identify potential concerns and opportunities.

Dispersion of metals arises from dissipative uses such as paints and fungicides, from corrosion of metals in harsh environments, from older incineration plants treating municipal or hospital waste, and from natural sources such as fossil fuels, fertilizers, soil erosion and volcanoes. Specific examples of dissipative uses include lead in gasoline additives, mercury in disposable batteries, cadmium yellow in artists' paint and copper in chemical sprays.

Metal dispersion has received a lot of attention in recent years and a number of metal uses have been discontinued. While banning of substances may seem an easy answer and is rational in a few cases it need careful assessment of costs and practical benefits, since alternatives may not exist, may be more costly, or less effective. Research is needed to objectively assess the reality of concerns and to explore whether alternative measures can provide adequate risk control.

The industry is working to get the facts needed to properly assess and manage the risks related to such dispersion. This includes quantifying sources, identifying species and assessing realistically the nature and magnitude of the ecological and health effects. Much more cooperation between the organizations involved in the different stages of metal treatment and in other industries which inadvertently disperse metals (power & agri-chemicals) is required. Greater engagement of the biological scientists in this research would be beneficial, particularly in assessing effects of the mining and metals industry on eco-systems.

6. Energy Efficiency & Carbon Dioxide Emission Reduction

Some metallurgical plants have the advantage of captive hydro-electric facilities but this still does not relieve them from pressures to reduce energy consumption. If the mines were not using the hydro-electric energy, in most cases the surplus energy would be used to displace fossil fuel elsewhere. The pressure for reduced carbon dioxide emission is not going to lessen in the short term although under at least one scenario it may one day be viewed as a virtue since an expansion of global biomass production and inventory would otherwise denude the atmosphere of carbon dioxide.

Industry in the interim has to do all it can to improve energy use efficiency. This in itself is good business provided regulations and taxes designed to accelerate carbon dioxide reduction are not imposed distorting normal competitive relationships. Such measures can, in a global perspective, be counter-productive by favoring less efficient off-shore producers in less responsible jurisdictions. Large amounts of energy are used in the extraction, transportation, crushing, grinding and separation of ore and all production stages invite research to develop improved techniques. The mining industry makes use of the energy in the sulphur in the ore for autogenous smelting to reduce fossil fuel consumption.. The industry facilitates technology transfer domestically through participation in a number of associations and internationally through construction of new facilities and through personnel education and training.

7. Biotechnology Applications in Metal Mining

The next few decades are going to see an explosive growth of biological knowledge. Mining applications are already evident in the leaching of metals, in the control of acid mine drainage, oxidation of thiosalts, remediation of soils and ground waters, the growth of plants, and the concentration of metals. Many more can be expected.

SUMMARY

Global View

A recent report of the United Nations Environmental Program identified four major global challenges that need to be addressed more vigorously. The mining industry is contributing beneficially in all four areas and needs to seek ways to do better. Table 6. Summarizes the challenges and mining responses.

Table 6: Key Concerns from UNEP Report
'Global Environmental Outlook for Life on Earth'

UNEP Concern	Mining Contribution
Energy Efficiency and Renewable Energy	**Promotion through CIPEC and the VCR and hydro installations in remote locations.**
Distribution of Environmental Technologies Worldwide	**Latest and best technology in new plants in developing countries**
Ensuring Long Term Availability of Fresh Water	**Recycling water cleaned to potable standards for beneficial uses**
Bench Marking Data and Integrated Assessments	**Implementing ISO 14000 EMS and Sustainable Development Policies**

Qualitative Benefits

The industry research program is driven by the anticipation that the investment in research will generate generous returns in the form of improved operations or reduced risks. Past experience can provide examples of benefits in each of the areas listed in Table 7.

Table 7: Examples of Qualitative Justifications for Specific Research Programs

General Research Goal	Research Benefit Examples
cleaner technology	lower losses/better people
best available technology	faster approvals/greater profits
niche market development	extra revenues/cut cover costs
improved stockpile isolation	visual credits/less on-going cost
life cycle management	better image/more recycle metal
more efficient energy use	less CO_2/lower energy costs
enlightened risk assessment	fewer upsets/less insurance

Cooperative Research
There is a need for much greater degree of cooperation across industry in the field of research to get maximum return from limited funds. Opportunities are shown in Table 8.

Table 8: Opportunities for Cooperation on Research Projects.

Organization	Benefit
Greater use of industrial consortia to working on common concerns.	Costs and capabilities shared with synergistic benefits.
Promotion of Centres of Excellence with broad objectives	Get benefit of cross fertilization of ideas from many academic perspectives.
Establishment of virtual research institutions.	Centre of Excellence with an explicit budget and mission linked by Internet
Encouragement of work shops to promote synergistic thinking.	Directed to making products of research available to actual users.
Facilitation of government funding of appropriate university research.	Using seed money to tap into NSERC or other funds on a small project basis.
Involvement of biologists to study mine related biological systems.	Moving from classic engineering disciplines to link in bio-technology.
Intense fundamental studying possible links between rock microstructures.	Seeking leads to new and improved rock breaking and grinding procedures
Traditional applied research with external reviews.	Focused teams on applied research in company laboratories.
Continual process improvement with external links.	Operations centred/plant focused investigation of existing facilities.

CONCLUSIONS

Future Research Organization

Research responsibility has been traditionally divided between company owned and operated research facilities striving for major break-throughs and in plant technical groups working on incremental improvement of existing processes. This is gradually changing. The dividing line between the two groups is becoming blurred as inter-change of personnel increases. Further electronic communication is facilitating the functioning of a wide variety of other organizational structures including virtual universities and Centres of Excellence.

Cooperative Research Need

My fellow panelists have laid out some of the challenges and suggested some creative ways in which we can improve industry performance. I think that they have been too kind to the industry. While much has been accomplished much more remains to be done and only through the collective, constructive, innovative and intensive efforts of the entire metals producing, fabricating, manufacturing, using and recycling industry, in conjunction with consumers and municipalities, can the level of performance that society desires and demands be achieved.

Improved Research Management

Research has contributed materially to the economic and environmental improvement of the metals industry in the past and has an even more important role to play in the future. The industry is highly competitive, the funds available for research are limited while there is a need for a greater, more productive and more diversified research contribution. The research manager has to find more effective ways of meeting all these conflicting demands in a cost effective way.

Call To Action

The technical and economic restructuring of the metal mining and processing industry through research is an exciting challenge and we can expect many interesting and profound developments to flow out of this effort over the next decade. The challenge is great and for the full potential of research to be realized the entire mining and metals industry has to acknowledge the magnitude of the challenges facing us and cooperate in a more efficient and effective way than in the recent past. The rate of expenditure on research by industry needs to double to keep ahead of the challenges facing us. This expenditure will be amply rewarded in terms of more efficient operations.

The Basel convention and pollution prevention: Are these appropriate tools to meet a sustainable development objective?

G.B. Smith
Minerals and Metals Sector, Mineral and Metal Policy Branch, Nonferrous Division
Department of Natural Resources
580 Booth St., Ottawa, Canada K1A 0E4

ABSTRACT

The Basel Convention is a multilateral environmental agreement designed to reduce the transboundary movements of hazardous wastes and, in particular, the dumping of hazardous wastes in developing countries. The Convention's most important commitment is to control the transboundary movement of "hazardous wastes" in an "environmentally sound manner". The instruments of control are provided through a Prior Informed Consent procedure, which requires formal descriptive confirmations of pending shipments and notification of governments representing the site of destination. The underlying objective is for countries to assume a responsibility for the hazardous wastes they generate which indirectly may stimulate the use of cleaner production methods thereby reducing the generation of hazardous wastes.

Pollution Prevention is designed to focus upon upstream production processes to reduce or eliminate the generation of pollutants and wastes thereby eliminating the need to manage these substances at the end of the tailpipe.

Any evaluation of the relationship between these two instruments and Sustainable Development requires a clear understanding of what is a pollutant and a waste in the context of Pollution Prevention and what is a hazardous waste in the context of the Basel Convention.

Proceedings of
Metals and the Environment: An International Symposium
Edited by Jim Skeaff
The Metallurgical Society of CIM
100th CIM Annual General Meeting, 1998

BASEL CONVENTION

The Basel Convention has suffered from an inability to precisely define which materials are considered as "hazardous wastes". The Convention defines "wastes" as being "substances or objects which are disposed of or are intended to be disposed of or are required to be disposed of by the provisions of national law". The Convention attempts to qualify the scope of "wastes" by reference to an Annex 1 which lists "waste streams" arising from various sources or operations and "wastes having as constituents" certain listed elements. As the metallurgical industry is mature, there exists significant international recycling of minerals and metals. Accordingly, many of these inorganic recyclables fall within the scope of the Basel Convention. The Convention also includes within its scope specific non-hazardous wastes which must be managed equivalently to hazardous wastes. The Convention defines "hazardous" by reference to an Annex III which lists a series of acute and chronic hazard characteristics. The acute hazards are generally taken from the United Nations Transport of Dangerous Goods Guidelines and, as such, enjoy a significant measure of international recognition supported by harmonized test protocols. The chronic hazards are poorly defined, are not supported by harmonized international test protocols and enjoy mixed acceptance in use internationally particularly when they are applied to recyclable materials.

In defining "wastes" as those materials to be disposed of, the Basel Convention has further been plagued by their decision to include the economic activity of "recycling" within their list of "disposal operations" contained in an Annex IV.

The Basel Convention was signed in 1989 and entered into force in 1992. As of January 1998, 112 countries have ratified this Convention with the most notable exception being the United States. Since its entering into force, the Convention has devoted significant amounts of time, energy and resources to clarify which materials actually fall within the scope of the regime. The need for clarification has been exacerbated by an amendment adopted by consensus in 1995 proposing to ban the export of all hazardous recyclables arising from ostensibly OECD countries and destined for non-OECD countries by 31 December 1997.

To this end, a Basel Convention Technical Working Group (TWG) has developed a list of wastes and waste streams that provide a level of improved clarity for practical implementation of a "ban" amendment. This "hazardous waste list" is proposed to be included into the Convention via an amendment to be discussed at the fourth Conference of Parties scheduled for February 1998. While this negotiated hazardous waste list should certainly improve the clarity of scope of the Convention, there remain recyclable materials contained on this list that are considered as essential raw materials for use by recycling industries and which, because they exhibit a hazard characteristic, will be subject to the movement "ban" regime.

Despite the significant work conducted by the Basel TWG, many Annex VII countries have stated their intention to implement the "export ban" based on their domestic determination of 'hazardous waste". Accordingly, the scope of "banned" substances will be significantly broader than that proposed by the Basel Convention.

POLLUTION PREVENTION

Canada adopted a Federal definition of Pollution Prevention (PP) in 1995. Pollution prevention, as defined by Canada, is "the use of processes, practices, materials, products or energy that avoid or minimize the creation of pollutants and waste, and reduce overall risk to human health or the environment".

Pollution prevention targets pollutants and wastes for reduction in generation. While a precise definition of "pollutants" and "wastes" was not specifically included in this program, there is a discussion related to recycling practices and processes as they relate to pollution prevention.

The Canadian Pollution Prevention Strategy for Action notes that

> "recycling is both an environmental protection method and a business. On-site, recycling of materials is considered to be a pollution prevention practice. When destined for recycling, materials of commercial value are implicitly not considered wastes and pollutants. Off-site recycling is considered to be a separate and much encouraged business activity."

SUSTAINABLE DEVELOPMENT

The goal of sustainable development (SD) is to "meet the needs of the present without compromising the ability of future generations to meet their own needs".

This definition of SD arose from the World Commission on Environment and Development (the Brundtland Commission) and has been accepted by the Government of Canada. The Government has built upon this definition of SD as a point of departure in applying the concept to minerals and metals. Canada's Policy, the Minerals and Metals Policy, incorporates the following elements derived from an SD objective:

- finding, extracting, producing, adding value to, using, re-using, recycling and, when necessary, disposing of mineral and metal products in the most efficient, competitive and environmentally responsible manner possible, utilizing best practices;

- respecting the needs and values of all resource users, and considering those needs and values in decision-making;

- maintaining or enhancing the quality of life and the environment for present and future generations; and

- securing the involvement and participation of stakeholders, individuals and communities in decision-making.

Importantly, defining SD in the context of minerals and metals, it should be recognized that the economic and social benefits of mineral development are not all consumed by the present generation.

The fabric of SD is woven together with the three cords of social responsibility, environmental protection and economic development. Achieving SD, requires that environmental, economic and social considerations must be fully taken into account as early as possible in the decision-making process.

Additionally, in applying the philosophy of SD to minerals and metals, the Government of Canada has developed the "Safe Use Principle". This Principle calls for an assessment of the risks at various life cycle stages associated with the uses of mineral or metal containing products. These risk assessments can then be used to determine an appropriate management approach to address these risks. In practical terms, where the risks exhibited by a particular use of a mineral or metal containing product cannot be adequately managed, it may be appropriate to discontinue this specific product use.

With world population growth and rising standards of living, global demands for minerals and metals as essential raw materials for economic and social development continues to increase. This growth in demand presents a range of policy issues concerning sustainable use of minerals and metals and their availability for future generations.

There are a number of factors that will promote the continued availability of mineral and metal resources. An important component in this continued availability is the use of recycled materials. Because of their value, consistent performance characteristics, durability, chemical properties and versatility of use, many minerals and metals can be re-used almost without limit. As a consequence, recycled materials are a vital secondary source of supply, and are traded on national and global markets.

As such, it is important to promote and enhance the increased use of recyclable metals as a means of ensuring their availability for future generations.

COMPATIBILITY OF THE BASEL CONVENTION WITH SUSTAINABLE DEVELOPMENT

The question raised within the context of minerals and metals use relates to whether the Basel Convention and Pollution Prevention are compatible with the philosophy of SD?

Before considering the question of compatibility, it is important to recognize that both the Basel Convention and PP are environmental actions. As such, their primary objective is to achieve a measure of environmental protection. In that SD requires a harmony or balance among environmental actions, economic activity and social welfare, environmental actions must be evaluated not only on whether they achieve their environmental objectives but whether, in achieving these objectives, they have appropriately considered the economic impacts and social aspects of the SD equation.

As a guiding principle for the use of metals and minerals, the path towards SD should point to the wise and efficient use of resources within an environmentally sound management infrastructure. As a rule of thumb, programs that severely restrict the use or significantly impede the access to and movement of basic resources should be considered suspect and would appear to conflict with the objectives of SD. Equally, programs that disregard potential environmental risks and fail to incorporate appropriate environmental protection component elements will generally conflict with SD goals.

The pre-1994 Basel Convention, (pre export ban Decision II/12 and pre export ban amendment III/1) may appear to be incompatible with the principles of SD in that the Convention fails to recognize the economic and social benefit potential that can be derived from the safe use of recyclable resources. In that the Basel Convention defines "recycling" to be a "disposal operation', it creates a significant negative stigma associated with all recycling operations and the continued use of recyclable resources. As such, the Convention does little to promote environmentally sound recycling opportunities with particular attention to the significant economic, social and environmental benefits that could accrue to developing countries. The Convention casts recycling in the negative light of being considered equivalent to a hazardous waste final disposal activity.

The negative stigma associated with the terms "waste" and "disposal" have direct implications for use, distribution and resource management which lead to a reluctance to recycle and an unwillingness to use products arising from recycling processes. It is the view of the Government of Canada that the term "waste" should be reserved for those materials destined for final disposal operations.

While governments collectively negotiated and agreed to the definitions contained within the Basel Convention, they alone should not be held solely responsible for the error of categorizing recycling as a waste disposal activity. While the international metallurgical industries have long recognized the importance of the use of recycled metals, they have failed to sufficiently communicate the importance of recycling nor have they adequately emphasized the significant benefits accrued from environmentally sound recycling activities to governments and to international communities. The result of this lack of information is that world governments have individually and collectively defined a class of material as "waste" that contributes to 50% of the world's production of steel, 50% of the world's production of lead, 40% of the world's production of copper, 30% of the world's production of aluminum and 25% of the world's production of zinc.

Putting aside, however, Basel's definitional inconsistency with the philosophy of SD, the pre-1984 Basel Convention served to accomplish a specific environmental goal; that being to eliminate the dumping of hazardous waste in countries least capable of managing them and to introduce a movement control regime for these materials that would allow countries the opportunity to agree or disagree to their entry into their jurisdictions. While some of the provisions of the Basel movement control regime may appear onerous and, for some materials, unjustified, all countries were able to access the recyclable resources for which there existed an economic need. As such, the case can be made that the pre-1994 Basel Convention was generally consistent with the objectives of SD.

There is, however, a major difference between a movement control regime and an export ban. With the adoption at the 1994 second Conference of Parties, of the Decision II/12 banning the export of hazardous recyclables arising from OECD countries and destined for non-OECD countries, the Basel Convention completely turned away from SD and became a trade-controlling "club" comprised of OECD countries. This mutation was complete with the adoption by consensus of the 1995 "ban" amendment at the third Conference of Parties. While the "ban" amendment language specific to OECD was removed and replaced by an Annex of countries, the OECD/non-OECD distinctions remain.

Movement bans are considered punitive in nature and raise a large potential for inappropriate application and resource access restrictions using environmental protection as a guise for imposing non-tariff barriers to trade. In the context of an environmental action, "bans" are considered a measure of last resort to accomplish specific environmental objectives when efforts at all other less onerous avenues have been exhausted. It is the policy of the

Government of Canada to promote the elimination of unjustifiable non-tariff barriers that impede the industry's access to international markets.

While the original Convention included the ability of countries to impose a "ban" on the movement of specific recyclables, this "ban" was an "import" ban which allowed country's the ability to ban, of their own accord, the import of recyclables into their sovereign jurisdiction. The 1994 ban Decision and the 1995 "ban" amendment are export bans which allow exporting countries the ability to impose access restrictions on potentially essential raw material feed supplies.

This breach from the principles of SD has been widened by the actions of many OECD countries who have unilaterally decided to exclude the possibility for non-Annex VII countries to establish bilateral agreements in order to access some specific recyclable materials for which there may exist an essential need. Many of these same OECD countries are proposing action designed to strengthen this movement ban by eliminating the possibility for non-Annex VII countries to become members of Annex VII thereby gaining access to some needed raw materials. Clearly these collective trade actions are self serving and resource protectionist in nature. Such measures are contrary to the objectives of the Basel Convention whose goal is the promotion of environmentally sound management.

The Basel Convention "ban" amendment will, by its design, impact developing countries whose resource needs are most significant. The demanding resource requirements of developing countries, particularly those undergoing a state of industrialization, their increasing developmental needs and either their lack of naturally occurring mineral and metal resources or their lack of financing to access these natural resources will leave many industrializing non-OECD countries with few resource choices for building developmental infrastructures. The Basel Convention's disregard for the economic developmental needs of developing world economies should be corrected.

Until such time as recyclable materials are properly considered as resources and access to these materials is permitted to any country that declares or can demonstrate their ability to manage the materials in an environmentally sound manner, the Basel Convention will not be considered consistent with the principles of Sustainable Development.

COMPATIBILITY OF POLLUTION PREVENTION WITH SUSTAINABLE DEVELOPMENT

It is my view that PP as defined within Canada is generally consistent with the goals of SD. The Canadian Federal PP program distinguishes between materials that generate and earn revenue for businesses and those materials which industry is required to pay to eliminate and, accordingly, drain industry resources. It is evident that materials requiring expenditures by industry to mange are clearly those which industry themselves would prefer not to generate.

On reflection, this makes practical sense. It is logical that materials of value will be managed in a manner reflective of their economic nature. Generators of valuable materials do not dispose of these assets indiscriminately. They sell these assets into markets that equally value receiving the assets in full and good condition. Equivalent incentives exist for both seller and buyer. A shipper sells the good and a receiver purchases the good. An incentive exists on the receiver's part to ensure receipt of all the good as he is prepared to pay for the good. The shipper has an incentive to ensure all the good is delivered as he anticipates being paid for the good. Both parties will work to ensure all of the good is delivered and received.

There may be, however, quite different incentives involved in the movement of wastes, hazardous wastes and recyclable materials of negative economic value. In this scenario, the shipper pays the receiver to accept the material. The receiver, who has agreed to accept this material for a fee, recognizes that its treatment or disposal will cost money.

In this instance, if all the material doesn't arrive at the receiving facility, the shipper pays less. Equally, the receiver has an incentive to use the most cost effective treatment option for this material. If not all the material is subject to the disposal or treatment process, treatment or disposal costs may be reduced. This creates a potential incentive for mismanagement.

The Canadian PP exclusion of recyclable materials that are sold into the marketplace serves as an incentive to reduce or eliminate the generation of materials that industry neither wants to generate nor manage. There is a built-in incentive to review production process emissions, effluents and solid wastes to see if modifications to these streams can result in their transformation through production to become value added-products suitable for marketing.

PP includes on-site recycling of wastes and pollutants as an appropriate PP tool. In that the material is re-routed back into an on-site production stream, the intent is to re-use or valuate this substance through re-introduction into a commercial application.

Importantly, while PP does not include "off-site recycling" within its "tool-box" of activities for wastes and/or pollutants, the PP philosophy acknowledges that off-site recycling of pollutants and wastes is a responsible and desirable management option, preferred to final disposal alternatives. PP advocates, however, that the preferred and promoted objective is not to generate the waste or pollutant in the first place. Inherently, industry will agree that eliminating the generation of these substances is a preferred outcome.

Accordingly, pollution prevention can legitimately be considered as promoting the objectives of SD. Pollution prevention actions such as eliminating the generation of wastes and pollutants, encouraging facility operation and process reviews with the goal to improving management capabilities, fostering process design changes to aid environmental management, implementing product redesign and product reformulation and promoting the use of less toxic materials in cases where the risks of exposure and release are not sufficiently manageable are all considered consistent with the principles of SD.

Needs and options: Broadening the environmental management tool box

Jennifer Moore
Director General, Regulatory and Economic Issues, Policy and Communications
Environment Canada
Place Vincent Massey, 351 St. Joseph Blvd., Hull, Québec, Canada K1A 0H3

ABSTRACT

The environmental management tool box consists of a range of instruments from education and outreach, to standards and codes of practice, to voluntary agreements such as memoranda of understanding, to market-based instruments, and to regulations ranging from traditional command and control to performance-based.

The goal of government is to demonstrate measurable environmental results. This can be achieved by looking at the expanded tool box and engaging industry in meaningful dialogue. Examples of how this dialogue is shaping and broadening the tool box include: Whitehorse Mining Initiative, the federal Minerals and Metals Policy, update to the Metal Mining Liquid Effluent Regulations, the Strategic Options Process for the base metal smelting industry, the Accelerated Reduction and Elimination of Toxics (ARET) challenge, and the Voluntary Challenge and Register (VCR) to reduce greenhouse gases. The future challenge is to broaden these applications and explore other areas where these tools can result in enhanced environmental quality.

Proceedings of
Metals and the Environment: An International Symposium
Edited by Jim Skeaff
The Metallurgical Society of CIM
100th CIM Annual General Meeting, 1998

INTRODUCTION

The environmental management tool box consists of a range of instruments from education and outreach, to standards and codes of practice, to voluntary agreements such as memoranda of understanding, to market-based instruments, and to regulations ranging from traditional command and control to performance-based.

All levels of government are looking to expand their tool box by engaging in a more meaningful dialogue with industry. From an environmental perspective, our goal is to demonstrate measurable environmental results through new partnerships with all sectors of society. The recent work of the New Directions Group to develop a framework of criteria and principles for parties interested in the expanded tool box for environmental protection will provide a strong basis for future consideration of these tools. Some of these criteria include: certainty, flexibility, transparency, accountability, measurable targets and dates, improved economic efficiency, competitiveness and social equity.

NEED FOR REGULATORY FLEXIBILITY TO STIMULATE INNOVATION AND NEW TECHNOLOGIES

These "new" tools are complementary to regulations. They will allow for greater flexibility in achieving environmental objectives, and at the same time stimulate innovation and new technologies. Governments need to be open to the use of these tools in appropriate situations, but they should also ensure that the necessary regulatory backstops are in place to ensure that there is a level playing field amongst all participants for a particular initiative.

CONSULTATIVE APPROACH

There are some good examples of how ongoing dialogue is shaping and broadening the tool box. The Accord from the September 1994 Whitehorse Mining Initiative was the first attempt to articulate sustainable development in a policy statement for the natural resources area, and provide a framework for addressing concerns about the regulatory regime for mining. This vision was based on the consensus view of governments, industry, labour, selected communities and aboriginals. It articulated the broad objective of providing more efficient and effective regulation while maintaining high environmental standards. More recently, the federal Minerals and Metals Policy which was also based on extensive consultations, advanced the concepts of using a mix of regulations, non-regulatory approaches and market mechanisms as important means for achieving sustainable development.

REGULATORY REVIEW: ROLE OF SOUND SCIENCE

Clearly there is a need for regulations and standards that are based on sound science. The federal Metal Mining Liquid Effluent Regulations, promulgated in 1977, provide a framework for water quality improvements based on sound science, technology and economics. These regulations are currently being updated as a result of a multistakeholder review process, and will include an environmental effects monitoring program to measure environmental effectiveness. A consultative process (Strategic Options Process) has also been undertaken with base metal smelting firms to look at options for managing substances declared toxic under the Canadian Environmental Protection Act. Recommendations coming out of this process included voluntary reductions of releases, development of performance standards and site specific plans to enhance current regulatory requirements.

VOLUNTARY CO-OPERATIVE PROGRAMS

Participation in the Accelerated Reduction and Elimination of Toxics (ARET) challenge is starting to show results. More than 90 percent of the Mining Association of Canada's member companies are on track to reduce emissions by 75 percent for certain metals including zinc, lead and copper and hydrogen sulphide by the year 2000. As well, some mining firms are part of the Voluntary Challenge and Register to reduce greenhouse gas emissions. Post-Kyoto, as we develop national implementation strategies over the next two years, we will be looking at ways to strengthen and deepen these commitments, as all sectors of society work together to meet Canada's 6 percent emission reduction target.

INTERNATIONAL INITIATIVES

Other examples where work is ongoing are in the areas of ISO 14000 standards and the associated voluntary conformity assessment mechanisms such as registration of companies and the certification of environmental auditors. Linkages are also being established to international efforts in promoting eco-efficiency including the improvement in resource productivity in industrialized countries over the next several decades (factor 4 and factor 10). We are also looking at the work of the Netherlands and the use of covenants.

MINISTERIAL INITIATIVES

In the area of regulatory reform, Mines Ministers at their meeting in July 1997 discussed the need for further progress. They requested a horizontal review of regulations affecting the mining sector, with the objective of identifying opportunities to work more cooperatively among the various levels of government and industry to improve existing regulatory regimes and related decision-making processes. The review, currently underway, will result in a report and action plan to be submitted to Mines Ministers this fall. This is to be followed up with a referral to the Canadian Council of Ministers of the Environment (CCME) for discussion in 1999 and ultimately developing and implementing an appropriate action plan.

Authors' Index